*PREDICTING THE FUTURE*

The Darwin College Lectures

# Predicting the Future

*EDITED BY LEO HOWE AND ALAN WAIN*

CAMBRIDGE
UNIVERSITY PRESS

PUBLISHED BY THE PRESS SYNDICATE OF THE UNIVERSITY OF CAMBRIDGE
The Pitt Building, Trumpington Street, Cambridge, United Kingdom

CAMBRIDGE UNIVERSITY PRESS
The Edinburgh Building, Cambridge CB2 2RU, UK
40 West 20th Street, New York NY 10011–4211, USA
477 Williamstown Road, Port Melbourne, VIC 3207, Australia
Ruiz de Alarcón 13, 28014 Madrid, Spain
Dock House, The Waterfront, Cape Town 8001, South Africa

http://www.cambridge.org

First published 1993
Reprinted 1993
First paperback edition 2005

*A catalogue record for this book is available from the British Library*

ISBN 0 521 41323 0 hardback
ISBN 0 521 61974 2 paperback

# CONTENTS

v

# Introduction
# Predicting the future

*LEO HOWE*

Whether there is a future to predict is not a question many people care to think about too deeply. Recent attempts to elucidate the circumstances of the contemporary world have nonetheless brought forth the diagnosis that we have reached the end of history. Given that the past and the future cannot stand alone, one can read into this the implication that the future too is at an end.

I trust that I am not alone in finding this idea both unappealing and contrary to the facts. Clearly, we live in a changing world. Social, economic and political change is going on all around us at what appears to be a constantly accelerating pace. Some scholars contend that such changes have few fundamental effects on the daily realities of existence: *plus ça change, plus c'est la même chose*. Although we live in an apparently changing world, in fact we all now inhabit one and the same world. Since history, so the current grandiose claim goes, is the unfolding of competing and conflicting ideologies, and since the great conflicts are behind us – the West has triumphed over the East; capitalism has defeated socialism; class conflict no longer has force in the world; science has conquered religion and superstition – we are witnesses to the end of ideology and even the end of history itself. And the end of history means, of course, the end of the future; as the new future reveals itself, we will simply get more of what we have already!

Such argument seems to me both simplistic and short-sighted. It ignores the fact, for example, that the major political unit of the world system, the nation-state, is at a critical juncture in its history. Just at the time that it appears to dominate the political stage other supranational units that may well sound the death knell of nationalism are beginning to emerge. Nationalism itself, apparently such a rampant force in the world today, is in fact a relatively modern phenomenon.

The 'end of ideology' argument also fails to provide an accurate description of the position of science. For some, science is an antidote to ideology and has forever rid us of the four idols (of the tribe, the market place, the cave and the theatre) which Sir Francis Bacon identified as the principal impediments to human understanding. For others, it is precisely the fact that science appears to disenchant the world that has given rise to a host of new religious movements – some of which adopt state-of-the-art technology to broadcast their messages – and a resurgence in aggressive evangelism and fundamentalism. For yet others, science is not anti-ideological because it constitutes itself an ideology. On this reading science is not the acid which corrodes class conflict and irrationality, but the product of a class society and therefore also a source and justification of domination in society.

Whilst it is obvious, moreover, that the old Soviet bloc with its communist ideology and totalitarian regimes has succumbed to the West, it is not nearly so obvious, now the honeymoon period is over, that capitalism can produce for the eastern European countries the economic prosperity to go along with their new-found political freedom. Some ex-Soviet satellites may well find themselves joining the Third World rather than the First World. It may also be the case that the centre of gravity of the world economy is anyway shifting further and further to the Far East. The very sluggish growth rates in Europe and the United States now compare unfavourably with those in Japan, parts of Southeast Asia and Southwest China. This may still be capitalism but it will look steadily less benign to the western world in the future if the West is no longer in complete control of it.

At least two further forces are implicated in producing what may be an uncertain and even radically different future: global environmental change and degradation, and the increasing poverty of the majority of the world's population. Whether science and technology – in the service of capitalism or vice versa – can generate remedies to counter these forces remains to be seen. It is anyway just as easy and as reasonable to argue that it is captitalism, industrialisation and technology which have themselves unleashed the forces that now threaten our very existence.

In short, we do still have a future, and in some senses it is likely to be far more complex than anything that has gone before. Accurate prediction and careful planning would therefore appear to be more necessary than ever. Yet prediction, as the essays in this volume amply demonstrate, is never straightforward and unproblematic. Stephen Hawking, for example, is able to identify two possible lines of development for the universe as a whole, the choice between the two depending on the total amount of matter it contains. The universe will either continue to expand to infinity or it will slow down, begin to contract, and end in the so-called 'Big Crunch'. But all this will take place over billions of years.

Ian Stewart, considering the nature and effects of chaos, and Frank Hahn, reflecting candidly on the role of prediction in economics, reveal that both nature and society are so complicated that to predict the weather or certain aspects of the economy for even a short time ahead is fraught with difficulty. The contradiction here is, however, more apparent than real for theoretical physicists may be able to plot the future course of the universe better than they can predict, because of the uncertainty principle, the outcome of single sub-atomic events happening in the next few microseconds. Economists, similarly, are better at short-range forecasting than long-range prediction.

Prediction and planning are not, moreover, neutral processes but ideological ones. Prediction is rarely done for its own sake; it is almost always prediction undertaken for some specific reason, and carried out by middle-class pundits, academics and politicians. What is predicted and how prediction is done is a deeply political process, as many of the contributions to this book clearly demonstrate. The issues those in power want forecasts about are not very likely to correspond to what the disadvantaged are most concerned with, or the pressures that most affect them in their daily lives.

Prediction is not therefore a simple concept, especially when one has the notion of time to incorporate. The nature and complexity of what one extrapolates from, the precision with which the processes of development are thought to be known, whether the outcome predicted has a contaminating effect on the prediction in question and may thus modify it, how far into the future this extrapolation is intended to predict, the range of variables which can be accommodated in calculations: all these are some of the many and more obvious problems which make foretelling the future a hazardous business.

Prediction is not a simple concept in a more serious and interesting manner also. How we predict, why we predict, who the predictors are, and what we expect of them: all to some extent depend, not just on the ideological and pragmatic concerns I have mentioned above, but on the kind of society we live in. A notion as fundamental as prediction does not exist in a vacuum. It is attached to a variety of other concepts which are linked in a systematic way and constitute part of the very fabric of thought of the society in question. What we mean by prediction is grounded therefore in a set of cultural assumptions about the relationship of the present to the past as well as to the future, about what we take to be knowledge about the world and how we arrive at it, and about how we conceive our environment, how we act in it, and how it acts on us. Ideas about the future will also be linked to related concepts of fate, providence, accident, free will, determinism, and so on.

Both through time and over space, societies differ in the relative concern they attach to the past, present and future. The nature of these differences has been hotly debated in the social sciences for many years. The influential French anthropologist Claude Lévi-Strauss drew the distinction between 'cold' and 'hot' societies. The former, the societies we used to disparage as 'primitive', are said to be characterised by a desperate resistance to any structural modification that would allow history to penetrate and change them; they are intent on preserving themselves by reproducing their past in the present, by keeping to traditional and customary ways of doing things, and by maintaining the distance between the present and the mythical past at a constant measure. Hot societies, on the other hand, developing for the first time during the Neolithic, are characterised by large-scale differentiation into classes and other social and economic categories. It is the conflict between these that generates energy and change. Clearly, such a distinction recalls Marx and Engels's famous claim, in the *Communist Manifesto*, that the 'history of all hitherto existing society is the history of class struggles'.

Similar distinctions, such as those between 'oral' and 'literate' societies, or between those societies based on myth and those based on history, have been advanced by other scholars. The oral/mythic societies, hardly different from 'cold' societies, emphasise the present, recapitulate the past in the present, and attempt to project the present into the future. The literate/historical societies, a sub-group of Lévi-Strauss's 'hot' societies, have print technology and can therefore store information in books. According to some, this innovation has enormous consequences which inexorably lead to modern, rational bureaucratic and scientific cultures. In these the present is a radical transformation of the past, and the future will be more than a simple extension of the present.

These newer formulations have some similarities with the grand evolutionary theories of the nineteenth century. Then, ideas about long-run historical change revolved about great transformations, for example, from magic through religion to science, from matriarchy to patriarchy, or from clan-based societies (with allegiance to kin and ancestors) to class-based societies (with allegiance to economic interests), or from *Gemeinschaft* to *Gesellschaft*, and so forth.

Schemes of this kind are not, of course, immune from attack and the critical literature surrounding them is huge. Whether such contrasts are interpreted as distinguishing two radically different types of society, or whether they should be taken to be of mainly theoretical interest – in which case the claim would be made that there is no concrete society which corresponds exactly to either type and that societies fall on a continuum between the two polar opposities – is an important issue but not one that needs to be explored here. The point to be stressed is that despite the exaggerated claims made by those advancing these evolutionary models, there are many instructive differences between societies in the way they conceptualise time past, time present and time future.

The process of predicting the future is not always the same. In different societies and at different times it has meant different things, and the notion of prediction has a history of its own. We did not always predict from the same assumptions as we do now, nor for the same reasons. Today accredited experts and specialists reveal the future of the economy, the natural environment, medical treatment, even the probable future of the universe, on the basis of scientific theory and empirical observation. In the past, soothsayers, priests, oracles and comets foretold the future on the basis of religious ideology and traditional authority. In so-called primitive societies shamans, diviners and ritual experts do not so much tell the future as attempt to reorder the present to bring it into line with the past. It is not that change and innovation do not occur, only that they are not recognised as changes because they are reinterpreted as novel manifestations of the already known.

One reason for such palpable differences is that 'modern', western society sees its origin in a quite different way to societies elsewhere in the world. From our scientific vantage point we see ourselves as ascending from lower forms of life, as being the beneficiaries of eons of evolution and material and technical progress; even the idea of moral progress probably still exists in the popular imagination, ditched though it is by the intelligentsia. We think of ourselves as progressing in a linear and dynamic fashion from a known past to an unknown future. And in the post-modern, post-structural era many would say, despite the booming heritage industry, that we live in a future-oriented society, one in which the future is more important to us than the past.

Other societies have a rather different view of things. The common factor is generally that they see themselves as the descendants of gods and mythical heroes, the present era being but a pale and tortured image of what was once a golden age. As Richard Gombrich reminds us, the great traditions of Hinduism and Buddhism are based on huge temporal cycles of decay and degeneration which will end respectively either with the destruction of the world or with the coming of a new Buddha, then to be followed by the resurrection of a pristine society.

But there is also uniformity between these societies and religious traditions for all of them espouse some form of return to a beginning – a theology of renewal and regeneration. Certainly this image is not alien to Christianity even if literal notions of the Second Coming and the Last Judgement no longer have much currency beyond rather narrow fundamentalist circles. The perpetual postponement of the Last Judgement does not constitute evidence of the invalidity of the belief. As Don Cupitt explains, the point is rather that these ideas function as admonitory pictures and guiding ideals which provide the framework for moral existence. Yet while this view of religion enables Cupitt to embrace linear time and an open future, Hawking asks us to accept that one distinct possibility for the future of the universe is that it will return to its beginning – a cycle of truly cosmic proportions!

# 1

## The future of the universe

*STEPHEN HAWKING*

This chapter is about the future of the universe, or rather what scientists think the future will be. Of course, predicting the future is very difficult. I once thought I would have liked to have written a book called *Yesterday's tomorrow: a history of the future*. It would have been a history of predictions of the future, nearly all of which have been very wide of the mark. But I doubt if it would have sold as well as my history of the past.

Foretelling the future in antiquity was the job of oracles or sybils. These were often women who would be put into a trance by some drug or by breathing the fumes from a volcanic vent. Their ravings would then be interpreted by the surrounding priests. The real skill lay in the interpretation. The famous oracle at Delphi in Greece was notorious for hedging its bets or being ambiguous. When the Spartans asked what would happen when the Persians attacked Greece, the oracle replied, 'Either Sparta will be destroyed, or its king will be killed.' I suppose the priests reckoned that if neither of these eventualities actually happened, the Spartans would be so grateful to the god of the sanctuary, Apollo, that they would overlook the fact that his oracle had been wrong. In fact, the king was killed defending the pass at Thermopylai in an action that saved Sparta and led to the ultimate defeat of the Persians.

On another occasion Croesus, king of Lydia, the richest man in the world, asked the oracle what would happen if he invaded Persia. The answer was that a great kingdom would fall. Croesus thought this meant the Persian Empire, but it was in fact his own kingdom that fell. He himself ended up on a pyre about to be burnt alive.

Recent prophets of doom have been more ready to stick their necks out by setting definite dates for the end of the world. These have tended to depress the stock market, though it beats me why the end of the world should make one want to sell shares for money. Presumably you can't take either with you.

A number of dates have been set for the end of the world. So far they have all passed without incident. But the prophets have often had an explanation of their apparent failures. For example, William Miller, the founder of the Seventh Day Adventists, predicted that the Second Coming would occur between 21 March 1843 and 21 March 1844. When nothing happened, the date was revised to 22 October 1844. When that date too passed without apparent incident, a new interpretation was put forward. According to this, 1844 was the start of the Second Coming. But first, the names in the Book of Life had to be counted. Only then would the Day of Judgement come for those not in the book. Fortunately for the rest of us, the counting seems to be taking a long time.

Of course, scientific predictions may not be any more reliable than those of oracles or prophets: one only has to think of the example of weather forecasts. But there are certain situations in which we think that we can make reliable predictions, and the future of the universe on a very large scale is one of them.

Over the last three hundred years we have discovered the scientific laws that govern matter in all normal situations. We still do not know the exact laws that govern matter under very extreme conditions. These laws are important for understanding how the universe began. However, they do not affect the future evolution of the universe unless and until the universe recollapses to a high density state. In fact, it is a measure of how little these high energy laws affect the universe now that we have to spend large amounts of money building giant particle accelerators to test them.

## CHAOTIC MOTION

Even though we may know the relevant laws that govern the universe, we may not be able to use them to predict very far into the future. This is because the solutions to the equations of physics may exhibit a property known as chaos. What this means is that the equations may be unstable to a slight change in the starting conditions. Change the way a system is by a small amount at one time, and the later behaviour of the system may soon become completely different. If you slightly change the way you spin a roulette wheel, for example, you will change the number that comes up. It is practically impossible to predict the number that will come up. Otherwise physicists would make a fortune at casinos.

With unstable and chaotic systems, there is generally a certain timescale on which a small change in the initial state will grow into a change that is twice as big. In the case of the Earth's atmosphere this timescale is of the order of five days, about the time it takes for air to blow right round the world. One can make reasonably accurate weather forecasts for periods up to five days. But to predict the weather much further ahead would require both a very accurate knowledge of the present state of the atmosphere and an impossibly complicated calculation. There is no way that we can predict the weather six months ahead beyond giving the seasonal average.

We also know the basic laws that govern chemistry and biology. So, in principle, we ought to be able to determine how the brain works. But the equations that govern the brain almost certainly have chaotic behaviour in that a very small change in the initial state can lead to a very different outcome. Thus, in practice, we cannot predict human behaviour even though we know the equations that govern it. Science cannot predict the future of human society or even if it has any future. The danger is therefore that our power to damage or destroy the environment or each other is increasing much more rapidly than our wisdom to use this power.

Whatever happens on Earth, the rest of the universe will carry on regardless. It seems that the motion of the planets round the Sun is ultimately chaotic, though with a long timescale. This means that the errors in any prediction get bigger as time goes on. After a certain time it becomes impossible to predict the motion in detail. We can be fairly sure that the Earth will not have a close encounter with Venus for quite a long time. But we cannot be certain that small perturbations in the orbits could not add up to cause such an encounter a billion years from now.

The motion of the Sun and other stars around the galaxy, and of the galaxy in the local group of galaxies, are also chaotic. By contrast, the motion of the universe on very large scales seems to be uniform and not chaotic. We observe that other galaxies are moving away from us, and the further they are from us, the faster they are moving away. This means that the universe is expanding in our neighbourhood: the distances between different galaxies are increasing with time.

We also observe a background of microwave radiation coming from outer space. You can actually observe this radiation yourself by tuning your television to an empty channel. A few per cent of the flecks you see on the screen are due to microwaves from beyond the solar system. It is the same kind of radiation that you get in a microwave oven, but very much weaker: it would only raise food to 2.7 degrees above the absolute zero of temperature, so it is not much good for warming up your take-away pizza. This radiation is thought to be left over from a hot early stage of the universe. But the most remarkable thing about it is that the amount of radiation seems to be the same from every direction. This radiation has been measured very accurately by the Cosmic Background Explorer Satellite. It is found to be the same in every direction; any differences that are observed are consistent with the noise in the experiment. There is no evidence of any variation in the background with direction to a level of one part in ten thousand.

In ancient times people believed that the Earth was at the centre of the universe. They would therefore not have been surprised that the background was the same in every direction. However, since the time of Copernicus, we have been demoted to a minor planet, going round a very average star, on the outer edge of a typical galaxy, that is only

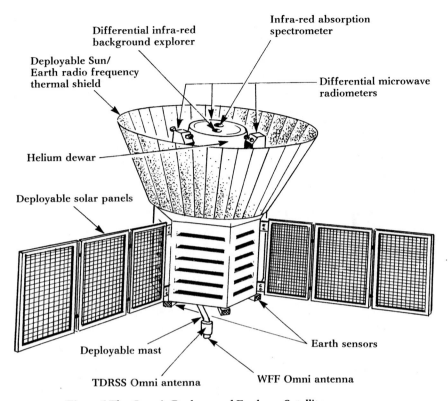

Differential infra-red
background explorer

Infra-red absorption
spectrometer

Deployable Sun/
Earth radio frequency
thermal shield

Differential microwave
radiometers

Helium dewar

Deployable solar panels

Deployable mast

Earth sensors

TDRSS Omni antenna

WFF Omni antenna

*Figure 1* The Cosmic Background Explorer Satellite

one of a hundred billion we can see. We are now so modest that we wouldn't claim any special position in the universe. We must therefore assume that the background is also the same in any direction about any other galaxy. This is possible only if the average density of the universe, and the rate of expansion, are the same everywhere. Any variation in the average density or the rate of expansion over a large region would cause the microwave background to be different in different directions. This means that on a very large scale the behaviour of the universe is simple and not chaotic. It can therefore be predicted far into the future.

### THE BIG CRUNCH

Because the expansion of the universe is so uniform, one can describe it in terms of a single number, the distance between two galaxies. This is increasing at the present time, but one would expect the gravitational attraction between different galaxies to be slowing down the rate of expansion. If the density of the universe is greater than a certain critical value, gravitational attraction will eventually stop the expansion and make the universe start to contract again. The universe would collapse to a Big Crunch. This would be rather like the Big Bang that began the universe. The Big Crunch would be what is called a singularity, a state of infinite density at which the laws of physics would break down. This means that even if there were events after the Big Crunch, what happened at them could not be predicted. But without a causal connection between events there is no meaningful way that one can say that one event happened after another. One might as well say that our universe came to an end at the Big Crunch, and that any events that occurred after were part of another, separate universe. It is a bit like reincarnation. What meaning can one give to the claim that a new baby is the same as someone who died, if the baby

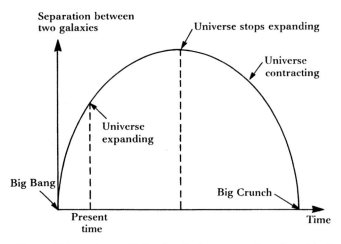

*Figure 2* The universe if the density is greater than the critical value

13

does not inherit any characteristics or memories from its previous life. One might as well say that it is a different individual.

If the average density of the universe is less than a critical value, it will not recollapse, but will continue to expand forever. After a certain time, the density will become so low that gravitational attraction will not have any significant effect on slowing down the expansion. The galaxies will continue to move apart at a constant speed.

So the crucial question for the future of the universe is, 'What is the average density?' If it is less than the critical value, the universe will expand forever. But if it is greater, the universe will recollapse and time itself will come to an end at the Big Crunch. I do, however, have certain advantages over other prophets of doom. Even if the universe is going to recollapse, I can confidently predict that it will not stop expanding for at least ten billion years. I don't expect to be around to be proved wrong.

We can try to estimate the average density of the universe from observations. If we count the stars we can see and add up their masses, we get less than one per cent of the critical density. Even if we add in the masses of the clouds of gas that we observe in the universe, it still only brings the total up to about one per cent of the critical value.

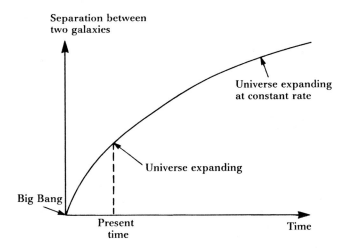

*Figure 3* The universe if the density is less than the critical value

However, we know that the universe must also contain what is called dark matter, matter that we cannot observe directly. One piece of evidence for this dark matter comes from spiral galaxies. These are enormous pancake-shaped collections of stars and gas. We observe that they are rotating about their centres. But the rate of rotation is sufficiently high that they would fly apart if they contained only the stars and gas that we observe. There must therefore be some unseen form of matter whose gravitational attraction is great enough to hold the galaxies together as they rotate.

Another piece of evidence for dark matter comes from clusters of galaxies. We observe that galaxies are not uniformly distributed throughout space, but that they are gathered together in clusters that range from a few galaxies to millions. Presumably these clusters are formed because the galaxies attract each other into groups. However, we can measure the speeds at which individual galaxies are moving in these clusters. We find they are so high that the clusters would fly apart unless they were held together by the gravitational attraction. The mass required is considerably greater than the masses of all the galaxies. This is the case even if we take the galaxies to have the masses required to hold themselves together as they rotate. It follows therefore that there must be extra dark matter present in clusters of galaxies outside the galaxies that we see.

One can make a fairly reliable estimate of the amount of the dark matter in galaxies and clusters for which we have definite evidence. But this estimate is still only about ten per cent of the critical density needed to cause the universe to collapse again. Thus, if one just went by the observational evidence, one would predict that the universe would continue to expand forever. After another five billion years or so, the Sun would reach the end of its nuclear fuel. It would swell up to become what is called a red giant, until it swallowed up the Earth and the other nearer planets. It would then settle down to be a white dwarf star, a few thousand miles across. So I am predicting the end of the world, but not just yet. I don't think this prediction will depress the stock market too much; there are one or two more immediate problems on the horizon. Anyway, by the time the Sun blows up we should have mastered the art of inter-stellar travel – if we have not already destroyed ourselves.

After ten billion years or so, most of the stars in the universe will have burnt out. Stars with masses like that of the Sun will become white dwarfs, or neutron stars, which are even smaller and more dense. More massive stars can become black holes, which are still smaller, and which have such a strong gravitational field that no light can escape. However, these remnants will still continue to go round the centre of our galaxy about once every hundred million years. Close encounters between the remnants will cause a few to be flung right out of the galaxy. The remainder will settle down to closer orbits about the centre, and will eventually collect together to form a giant black hole at the centre of the galaxy. Whatever the dark matter in galaxies and clusters is, it might also be expected to fall into these very large black holes.

One might expect, therefore, that most of the matter in galaxies and clusters would eventually end up in black holes. However, some time ago I discovered that black holes were not as black as they had been painted. The uncertainty principle of quantum mechanics says that particles cannot have both a well-defined position and a well-defined speed. The more accurately the position of a particle is defined, the less accurately its speed can be defined, and vice versa. If a particle is in a black hole, its position is well-defined to be within the black hole; this means that its speed cannot be exactly defined. It is therefore possible for the speed of the particle to be greater than the speed of light. This would enable it to escape from the black hole. Particles and radiation will thus slowly leak out of a black hole. A giant black hole at the centre of a galaxy would be millions of miles across so there would be a large uncertainty in the position of a particle inside it. The uncertainty in the particle's speed would therefore be small. This means that it would take a very long time for a particle to escape from the black hole; but it would escape eventually. A large black hole at the centre of a galaxy could take $10^{90}$ years to evaporate away and disappear completely. This is far longer than the present age of the universe, which is a mere $10^{10}$ years. Still, there will be plenty of time if the universe is going to expand forever.

## RECOLLAPSE

The future of a universe that expanded forever would be rather boring. But it is by no means certain that the universe will expand forever. We have definite evidence only for about a tenth of the density needed to cause the universe to recollapse. But there might be further kinds of dark matter that we have not detected which could raise the average density of the universe to the critical value, or above it. This additional dark matter would have to be located outside galaxies and clusters of galaxies. Otherwise we would have noticed its effect on the rotation of galaxies or the motions of galaxies in clusters.

Why should we think there might be enough dark matter to make the universe recollapse eventually? Why don't we just believe in the matter for which we have definite evidence? The reason is that having the critical density for the rate of expansion is very unstable. A small departure from the critical density in the early stages would have grown large as the universe expanded. In order that the density now should be within a factor of ten of the critical value, the initial density and rate of expansion would have had to be incredibly carefully chosen. If the density of the universe one second after the Big Bang had been greater by one part in a thousand billion, the universe would have recollapsed after ten years. On the other hand, if the density of the universe at that time had been less by the same amount, the universe would have been essentially empty since it was about ten years old. How was it that the initial density of the universe was chosen so carefully. Maybe there is some reason why the universe should have precisely the critical density?

There seem to be two possible explanations. One is the so-called anthropic principle, which can be paraphrased as, 'The universe is as it is because if it were different we wouldn't be here to observe it.' The idea is that there could be many different universes with different densities. Only those that were very close to the critical density would last for long and contain enough matter for stars and planets to form. Only in those universes will there be intelligent beings to ask the question, 'Why is the density so close to the critical density?' If this is the explanation of the present density of the universe, there is no reason

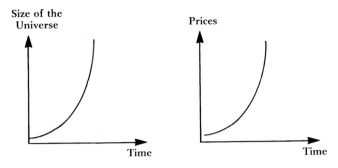

*Figure 4* Inflation

to believe that the universe contains more matter than we have already detected. A tenth of the critical density would be enough matter for galaxies and stars to form.

However, many people do not like the anthropic principle because it seems to attach too much importance to our own existence. There has thus been a search for another possible explanation of why the density should be so close to the critical value. This search led to the theory of inflation in the early universe. The idea was that the size of the universe might have kept doubling, rather as prices double every few months in some countries. However, the inflation of the universe would have been much more rapid and more extreme: an increase by a factor of at least a billion billion billion in a tiny fraction of a second. Of course, this was in the days before a Conservative government.

This amount of inflation would have smoothed out any irregularities or inhomogeneities in the very early universe. It would also have caused the rate of expansion of the universe to be so nearly the critical rate for its density that the rate of expansion and the density would still be near the critical relationship. Thus if the theory of inflation is correct, the universe must contain enough dark matter to bring the density up to nearly the critical value for the present rate of expansion. But because of the uncertainty principle of quantum mechanics, the universe could not be exactly the same everywhere and could not have exactly the critical density. There would have had to have been small fluctuations or uncertainties in the density and rate of expansion of the universe that varied from place to place. This means that the universe would probably recollapse eventually, but not for much longer than the fifteen billion years or so that it has already been expanding.

What could be the extra dark matter that must be there if the theory of inflation is correct? It seems that it is probably different from normal matter, the kind that makes up the stars and planets. We can calculate the amounts of various light elements that would have been produced in the hot early stages of the universe in the first three minutes after the Big Bang. The amounts of these light elements depend on the amount of normal matter in the universe. One can draw graphs with the amount of light elements shown vertically,

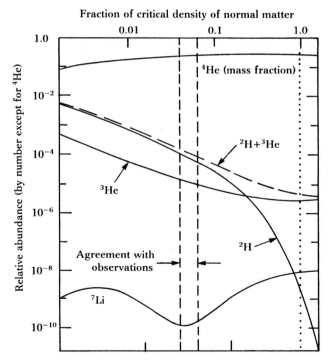

*Figure 5* The abundance of light elements formed in the early universe

and the amount of normal matter in the universe along the horizontal axis. One gets good agreement with the observed abundances if the total amount of normal matter is only about one-tenth of the critical amount now. It could be that these calculations are wrong, but the fact that we get the observed abundances for several different elements is quite impressive.

If there really is a critical density of dark matter, and it is not the kind of matter that stars and galaxies are made of, what could it be? The main candidates would be remnants left over from the early stages of the universe. One possibility is elementary particles. There are several hypothetical candidates, particles that we think might exist, but which we have not actually detected yet. The most promising case is a particle for which we have good evidence, the neutrino. Although this was thought to have no mass of its own, some recent observations have suggested that it does in fact have a small mass. If this is confirmed and found to be of the right value, neutrinos would provide enough mass to bring the density of the universe up to the critical value.

Another possibility is black holes. It is possible that the early universe underwent what is called a phase transition. The boiling or freezing of water are examples here: in a phase transition, an initially uniform medium like water develops irregularities like lumps of ice or bubbles of steam. These irregularities might collapse to form black holes. If the black holes were very small they would have evaporated by now because of the effects of the quantum mechanical uncertainty principle, as I described earlier. But if they were over a few billion tons (the mass of a mountain), they would still be around today and would be very difficult to detect.

The only way we could detect dark matter that was uniformly distributed throughout the universe would be by its effect on the expansion of the universe. One can determine how fast the expansion is slowing down by measuring the speed at which distant galaxies are moving away from us. The point is that we are observing these galaxies in the distant past, when light left them on its journey to us. One can plot a graph of the speed of the galaxies against their apparent

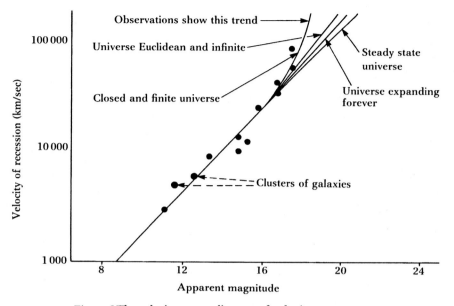

*Figure 6* The velocity versus distance of galaxies

brightness, or magnitude, which is a measure of their distance from us. Different lines on this graph correspond to different rates of slowing of the expansion. A graph that goes straight, or flattens out, corresponds to a universe that will expand forever. And a graph that bends up corresponds to a universe that will recollapse. At first sight the observations seem to indicate recollapse. But the trouble is that the apparent brightness of a galaxy is not a very good indication of its distance from us. Not only is there considerable variation in the intrinsic brightness of galaxies; there is also evidence that their brightness is varying with time. Since we do not know how much to allow for the evolution of brightness, we can't yet say what the rate of slowing down is: whether it is fast enough for the universe to recollapse eventually or whether it will continue to expand forever. That will have to wait until we develop better ways of measuring the distances of galaxies. But we can be sure that the rate of slowing down is not so rapid that the universe will collapse in the next few billion years. That should give us time to sort out a few immediate crises.

## THE POTENTIAL FOR TIME TRAVEL

Neither expanding forever, nor recollapsing in a hundred billion years or so, are very exciting prospects. Isn't there something we can do to make the future more interesting? An action that would certainly do that would be steering ourselves into a black hole. It would have to be a fairly big black hole, more than a million times the mass of the Sun. Otherwise the difference in the gravitational pull on one's head and one's feet would tear one into spaghetti before one got inside. But there is a good chance that there is a black hole that big at the centre of the galaxy.

We are not quite sure what happens inside a black hole. There are solutions of the equations of general relativity that would allow one to fall into a black hole and come out of a white hole somewhere else. A white hole is the time reverse of a black hole. It is an object that things can come out of, but nothing can fall into.

The white hole could be in another part of the universe. Thus this would seem to offer the possibility of rapid inter-galactic travel. The trouble is it might be too rapid: if travel through black holes were possible, there would seem nothing to prevent you arriving back before you set off. In theory you could then do something, like kill your mother, that would have prevented you from going in the first place. One only has to watch *Back to the Future* to see the problems that time-travel could cause.

However, perhaps fortunately for our survival and that of our mothers, it seems that the laws of physics do not allow such time-travel. There seems to be a Chronology Protection Agency that makes the world safe for historians by preventing travel into the past. What happens is that the uncertainty principle implies that spacetime is full of pairs of particles and antiparticles that appear together, move apart, and come back together again, and annihilate. These particles and antiparticles are said to be virtual because one does not normally notice their existence and one cannot observe them with a particle detector. However, if spacetime is warped so that particles can come back to earlier points in their histories, the density of virtual particles will go up because one could have many copies of a given particle at the same time. This extra density of virtual particles would either distort spacetime so much that it was not possible to go back in time, or it would cause spacetime to come to an end in a singularity, like the Big Bang or the Big Crunch. Either way, our past would be safe from evil-minded persons. The Chronology Protection Hypothesis is supported by some recent calculations that I and other people have done. But the best evidence that we have that time-travel is not possible, and never will be, is that we have not been invaded by hordes of tourists from the future.

## THE FUTURE OF THE UNIVERSE

Scientists believe that the universe is governed by well-defined laws that in principle allow one to predict the future. But the motion given by the laws is often chaotic. This means that a tiny change in the initial situation can lead to change in the subsequent behaviour that rapidly grows large. Thus, in practice, one can often predict accurately only a fairly short time into the future. However, the behaviour of the universe on a very large scale seems to be simple, and not chaotic. One can therefore predict whether the universe will expand forever, or whether it will recollapse eventually. This depends on the present density of the universe. In fact, the present density seems to be very close to the critical density that separates recollapse from indefinite expansion. If the theory of inflation is correct, the universe will actually be on the knife edge. So I'm in the well-established tradition of oracles and prophets of hedging my bets – by predicting both ways.

## FURTHER READING

Islam, J. N., *The ultimate fate of the universe*, Cambridge: Cambridge University Press, 1983.

Krauss, L., *The fifth essence*, London: Hutchinson Radius, 1989.

Lederman, L. M., and Schramm, D. N., *From quarks to the cosmos*, New York: W. H. Freeman, 1989.

Weinberg, S., *The first three minutes: a modern view of the origin of the universe*, London: Fontana, 1983.

# 2

## Chaos

*IAN STEWART*

> Let chaos storm!
> let cloud shapes swarm!
> I wait for form.
>
> Robert Frost, *Pertinax*

Mathematicians have always dreamed of explaining nature. However, their aspirations have often exceeded their abilities. The mathematician's standard mistake is to begin with a list of mathematical forms, derived from some general principle, and to force nature into the mould defined by those forms. For example, the ancient Greeks considered the straight line and the circle to be the most perfect forms. When the Renaissance mathematician Niccolo Fontana (nicknamed Tartaglia, 'the stammerer') described the flight of a cannonball, he constructed it from an initial straight line emerging from the mouth of the cannon, followed by an arc of a circle, followed by a vertical drop on to the target. Similarly, Johannes Kepler, seeking to explain not only the spacing of the planets but the reason for there being exactly six of them, devised a scheme whereby the five regular solids were neatly sandwiched between them.

Neither of these theories survived for long. Tartaglia's lines and circle were replaced by a parabola. To this day we have no very good theory of the distances between the planets, although an empirical law due to Titius of Wittemburg but usually credited to Johann Bode models them by a modified geometric progression. Kepler scored a palpable hit with his discovery that planetary orbits are elliptical. However, spectacular success was reserved for Isaac Newton, whose *Philosophiae naturalis principia mathematica* (1687) set out to discover the 'system of the world' – and did. It can be argued that Newton succeeded because he thought like a physicist rather than a mathematician. Instead of starting with preconceived ideas about what kind of mathematics nature ought to use, he derived his mathematical techniques from observations of what nature actually did. He thereby avoided the standard mistake, and created a paradigm for the mathematical modelling of nature, one so strong that it remained virtually unchallenged for three centuries.

Newton described – the current term is *modelled* – the behaviour of the physical universe in terms of mathematical equations, then known as 'differential equations' because they describe the difference between the present state and the state a small period of time into the future. Today a more impressive but also more vivid term is used: *dynamical systems*. In a dynamical system, the description of the near future is given purely in terms of the present state. So a dynamical system predicts behaviour a tiny instant into the future in terms of behaviour now. By repeating this procedure many times in succession, we can in principle predict the state of the system arbitrarily far into the future. Thus dynamical systems are *deterministic*, in a very strong sense: if they are put into exactly the same state twice, then they continue ever after in exactly the same way on both occasions.

According to the Newtonian paradigm, any system of bodies or particles is deterministic. In principle, this applies to the entire universe. As Pierre Simon de Laplace (1749–1827), a later exponent of the deterministic philosophy of mathematical modelling, put it:

> An intellect which at any given moment knew all the forces that animate nature and the mutual positions of the beings that comprise it, if this intellect were vast enough to submit its data to analysis, it could condense into a single formula the move-

ment of the greatest bodies of the universe and that of the
lightest atom: for such an intellect nothing could be uncertain,
and the future just like the past would be present before its eyes.

This is the *Hitchhiker's Guide* view of the universe: the super-
computer Deep Thought, computing the ultimate answer to the great
question of life, the universe, and everything. It is the paradigm of the
clockwork universe, never deviating from its initial course once its
cogwheels have been set in motion. And, for all its faults, it has been
spectacularly successful in helping humanity to come to terms with
the world around it. It does indeed encompass many aspects of life –
for example, the pattern of spots and stripes on the tail of a cheetah,
the rise and fall of populations of insects or fish, and the complex
folding of the DNA molecule. It captures much of the structure of the
universe, from Godfrey's Kinky Current, a huge and exceedingly
weird wind-formation found at the north pole of Saturn – it is an
almost perfect hexagon – through the spirals of galaxies, to the
moment of creation in the Big Bang.

## ... AND EVERYTHING?

Laplace thought mathematical models could explain all aspects of
our universe. However, science has a habit of emphasising its succes-
ses and sweeping its failures under the rug. Textbooks by the
thousand have been written explaining the triumphs of the deterministic view of the world; far fewer describe its shortcomings. The
dilemma is put into stark relief by a single picture, one of NASA's
*Voyager* photographs of Jupiter (Figure 1). The most prominent
feature is the Red Spot, remarkably stable amid Jupiter's swirling
storms, a whirlpool twice the diameter of the Earth. To human knowledge it has existed largely unchanged for at least three centuries,
ever since, as the *Philosophical Transactions of the Royal Society*
records, '... the ingenious Dr. [Robert] Hooke did some months since
intimate to a friend of his that he had, with an excellent 12 feet tele-
scope, observed, some days before he spoke of it (viz. on 1664 May 9)

*Figure 1* Order, Jupiter's Red Spot, spawns chaos, its turbulent wake

about nine o'clock at night, a spot in the largest of the three observed belts of Jupiter...' The Red Spot is stability and determinism incarnate. But, unseen until the *Voyagers* made their remarkable journey in 1979, the Red Spot is accompanied by cohorts of vortices, which spin off in a turbulent wake, much as eddies form behind the passing bulk of a whale. The vortices are irregular, patternless, ephemeral, unpredictable: entirely opposite characteristics to those of the Spot that spawns them.

The motion, and the stability, of the Red Spot are a consequence of the mathematical laws of fluid dynamics. These are deterministic equations, firmly in Newton's and Laplace's paradigm. But why is the stable, deterministic structure of the Red Spot accompanied by a random train of turbulent vortices? Do the same deterministic equations that explain the order of the Red Spot also govern the disorder of its wake? How can such a thing be possible?

The first theory of this phenomenon, and of turbulence in general, was that nothing of the kind is happening. As in all mathematical modelling of the real world, the laws of fluid dynamics involve certain approximations to the structure of a fluid – for example, that it is composed of infinitely divisible material, rather than discrete atoms. When turbulence occurs, it was widely held, then these approximations become invalid. The laws are different for turbulent flow. Different laws, different kinds of behaviour: what's the problem?

It was easy for this theory to remain unchallenged. The equations of fluid dynamics are complicated and intractable: when they can be solved by some mathematical formula, they necessarily behave in a nice manner. Turbulence is distinctly nasty, ergo, when a flow is turbulent, you cannot solve the equations – certainly not with pencil and paper, the traditional tools of the mathematician.

But now we have a new tool, the computer. It turns out that the same laws that govern the ordered flow of the Red Spot do indeed also govern turbulence. This is in agreement with the second theory of turbulent flow, credited to the Russian physicist Lev Landau (1908–68), in which the apparent irregularities are seen as tiny pieces of an overall order so complicated that we cannot perceive it as a whole. That theory is basically correct, although in the detailed mechanism that it proposed for this complex order – a mixture of infinitely many distinct periodic motions, like tiny vortices of wildly varying sizes – it is wrong. The current theory is that a new mathematical mechanism, dubbed *chaos*, is responsible. Chaos is simpler than Landau's picture, but far more subtle.

## BESTIARISING IS BEST

Progress comes if we take a mathematician's view of the whole problem. By focusing too firmly on turbulence, just one instance of irregular behaviour in an apparently deterministic system, we may be distracted by technical difficulties peculiar to that phenomenon. Instead we should take the 'basic science' view, asking simple but important general questions, without worrying about specific applications of them, without asking whether they are useful in some limited world-view. Once answers begin to emerge, however, we should remember to reinstate these more stringent criteria.

Medieval scholars liked to collect together descriptions of every living creature they could lay hands on, which they called bestiaries. Mathematicians like to build up bestiaries too, but the beasts within the mathematical pages are all of the things that can possibly happen in a given situation. Mathematical bestiaries are called classification theorems. The philosophy is that you do not understand a problem completely until you can bestiarise it – and even then you still may not understand it because there may remain unanswered questions about some items on the list. From this viewpoint, the crucial question is: What kinds of thing can a dynamical system do? Classical mathematics, for all its thousands of textbook examples, boils down to a very short list of general types of behaviour.

- A dynamical system can exist in a *steady state*, without changing at all. A good example is a rock, which sits for very long periods of time just being a rock, to the extent that every passing person remarks on the great symmetry and unchanging perfection of its steady state. Well, maybe not: whenever a rock suddenly changes its state we tend to notice; so in some sense we do perceive its ultimate steadiness.
- The next most complex behaviour is *periodicity*. A system is periodic if it keeps repeating the same behaviour: 'Christmas again – funny, it's only a year since the last one.' Astronomical cycles, such as the phases of the Moon or the turn of the seasons, are – to a high degree of approximation – periodic; so are many biological cycles such as breathing, sleep, or the heartbeat.
- The third type of motion that occurs in classical mathematics is called *quasiperiodicity*. The motion of the Moon round the Earth, on its own, is periodic, and the same is true for the Earth round the Sun. But what of the combined three-body system? If a whole number of lunar months were exactly equal to a whole number of years, then the combined Moon–Earth–Sun system would also be periodic, repeating exactly after those numbers of months and years. However, if no such whole number relationship exists – in which case the two separate periods are said to be incommensurable – then the motion is

29

quasiperiodic, or almost periodic. This is the ultimate among classically recognised motions. Since randomly chosen periods are in general incommensurable, the combined Moon–Earth–Sun system is presumably quasiperiodic.

The distillation of several centuries of mathematical analysis, often intricate and seldom easy, is that whenever the equations describing a physical system can be solved, then with very few exceptions indeed, the solutions are either steady, periodic, or quasiperiodic. From this point of view the universe appears to function as a vast system of superimposed cycles. The bestiary of classical dynamics contains only three beasts.

In the face of such overwhelming evidence, it takes a rather special attitude to question the perceived wisdom. In the end it was the pure mathematicians who asked: Can any other kind of motion occur? Pure mathematics has two characteristic (and often infuriating) features. It asks very general questions, and it requires logically watertight answers. And on this occasion the answer was surprising: there *do* exist other kinds of motion. The evidence only seemed overwhelming: in fact, it was a case of Catch 22. The unasked question was: What do the solutions look like when the equations can't be solved? Nobody knew, of course . . . because they couldn't solve them to find out!

## PHASE PORTRAITS

How do you find out what the solution to an equation looks like when you are not solving it? Obviously you have to approach the whole problem from some other direction. The mathematicians discovered that it is possible to visualise the solution, even if you can't write down a formula. This idea gained serious currency around 1900 in the astronomical investigations of the great French mathematician Henri Poincaré (1854–1912), and it led him to one of the first discoveries of what we now call chaos. Although Poincaré made little progress on chaos beyond the realisation that it can occur, he introduced a powerful new method for studying dynamical systems. It was based upon a new kind of geometry, which he called *analysis situs* but which we now know as topology: the qualitative geometry of the continuous.

Instead of focusing on formulas for their solution, Poincaré imagined a multidimensional space whose coordinates are the variables of the system. Each point in this space represents a whole set of coordinates, that is, a set of numbers determining the values of all of those variables. Each possible state of the system corresponds to a single point in phase space.

As time passes, the state changes, following the dynamical laws. The representative point moves, tracing out a curve in phase space. Solutions to the dynamical equations are represented by curves. A global picture of all possible solutions is given by collecting together all of these curves, for all possible starting values (in practice, a reasonable sample of them). This *phase portrait* captures in a single geometric picture not just what the system does but everything that it can do. It is often possible to determine the phase portrait without finding formulas for the solutions, so an overall qualitative feel for the dynamics can be gained by studying the geometry of the phase portrait.

Almost all of our understanding of dynamical systems has come through this geometric approach. For example, the Italian mathematician Vito Volterra (1860–1940) devised a simple model of the populations of predators and food fish in the Mediterranean sea. He was trying to explain a curious observation made by the biologist Umberto d'Ancona: during the First World War, when the amount of fishing decreased, the catches contained a greater percentage of predators. Volterra found the reason: the populations fluctuated periodically, in a manner that on average was to the predators' advantage. Figure 2 is a geometric representation of this discovery: the phase portrait of Volterra's equations. The solution curves are closed loops with a single point at their centre. This point represents a steady state solution, with prey and predators in perfect balance. The closed curves represent periodic solutions. The periodicity is

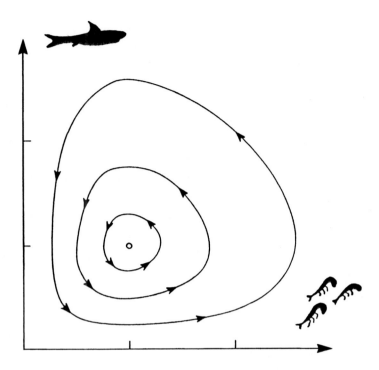

*Figure 2* The phase portrait of Volterra's equations for predator–
prey interaction. Closed loops correspond to periodic motion

essentially a consequence of delays in the response of the component populations. Imagine that the population of prey is high, but that of predators is low. Predators will then increase, depleting the food supply. Because of the relatively slow reproductive rate of predators, however, the population of predators will tend to overshoot the level that the food can sustain. Now the predator population drops, but until it becomes very low, the prey cannot increase substantially. Once there are few enough predators, however, there is a rapid explosion of the faster-reproducing food fish, and the cycle starts anew.

## DUCKING THE ISSUE

Volterra's system is an example of classical motion, the kind that Laplace would have approved of. What about chaos? Poincaré's astronomical researches are technically demanding, but fortunately we can get an excellent picture from something simpler. I will use a drawing of a duck to show that there is nothing very esoteric going on.

We've seen that the equations for a dynamical system effectively predict its future state – just a tiny instant into the future – in terms of the present one. They take the present state, and apply a fixed transformation to it to derive the state at the next instant. To predict the behaviour several instants into the future, we just repeat this procedure, a process known as iteration. Now transformations can be given by formulas; but they can also be given geometric descriptions. The formula $f(x,y)=(\frac{1}{2}x,\frac{1}{2}y)$, for instance, can be interpreted as 'shrink everything to half its size'. What happens if we iterate this? Figure 3a demonstrates the obvious with a volunteer duck. Everything shrinks, and shrinks, and shrinks, and eventually approaches as close as desired to a single point, called the fixed point. This fixed point is the geometric equivalent of a dynamical steady state: if you sit at the fixed point, then nothing happens. The duck collapses steadily around you, but you yourself don't move. The shrinking duck illustrates that no matter what the initial conditions may be, the long-term state is to remain at (or at least very close to) that single fixed point. There is no need to perform complicated calculations to see what would happen: it is immediately clear from the geometric picture.

Let us try another iteration, this time using the transformation 'rotate through a right angle'. As Figure 3b shows, the duck turns, turns, and turns again, a fourth turn bringing it back to exactly its original position. Thereafter, the motion repeats. So this time we get periodic motion, with a period of four time-steps. I could show you quasiperiodic motion in much the same way: rotate the duck through an angle that is not commensurable with 360°. It will keep coming

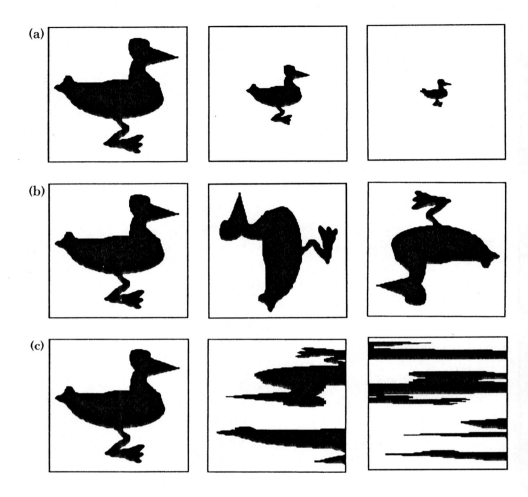

*Figure 3* Duck dynamics: (a) iterating a transformation that shrinks the plane – everything tends to a fixed point; (b) rotating the plane through a right angle leads to periodic motion; (c) kneading the plane mixes up duck and non-duck in an intricate *canard en croûte*

back to almost the same position, at almost regular intervals, but it will never get back exactly to where it started. But you can imagine that without a picture.

All of these types of motion are accessible to the classical formula-based approach, basically because the final result is fairly simple and

straightforward. But now let us try something a little more complicated: let us knead the plane like a baker making dough. Take a lump
of dough, stretch it out sideways, fold it over on itself, and repeat. An
idealised version of this transformation is to take a square in the
plane, stretch it out sideways to twice its width and half its height,
chop it in half down the middle, put the left-hand half back at the
bottom of the original square, and put the right-hand half at the top,
but upside down. Figure 3c shows what happens to our long-suffer-
ing duck: it becomes very long, and suffers almost total disintegra-
tion. You can imagine that after a few more iterations, the entire

square will appear a somewhat uniform grey, although if you looked very closely you'd see that it's composed of horizontal streaks of black and white interleaved like puff pastry – or in this case *canard en croûte*. Classical mathematics can easily write down a formula for such a transformation, but it can't solve the equations to discover the end result of iterating it many times.

This is chaos: highly complex, irregular, almost random behaviour occurring in a perfectly deterministic system. The revolutionary new discovery turns out to be something that cooks have known for centuries: repeated, regular kneading transformations mix everything up. The end result of this mixing process, dynamical puff pastry, is far too complicated for a simple formula to capture.

### GRAPHIC DISCOVERIES

Two major changes in the way mathematicians went about their job were crucial to the discovery of chaos. On the intellectual side it required a move away from formulas and towards geometrical thinking, a switch from quantities to qualities. But qualitative results alone are not always very useful: some quantitative substitute for the classical formulas had to be found. This problem was solved with new technology, the invention of fast computers with big memories and good graphics. With computers we can solve an equation by following through the procedure for unravelling the future states and listing the results as tables of numbers. Alternatively, for rapid comprehension we can draw them graphically.

Figure 4 shows some futures predicted by a mathematical model of the production of white blood cells. The model, based on the physiology of the process, assumes that the body detects a shortage of white cells in the blood and passes a chemical message to initiate production of more, from so-called stem cells. There is a time delay while the cells mature, and they are then released into the bloodstream. When the quantity in the bloodstream gets too large, the chemical signal is turned off and production from stem cells ceases. The regulating mechanism is rather like a thermostat, but one that is rather slow to operate because the time delay is quite large.

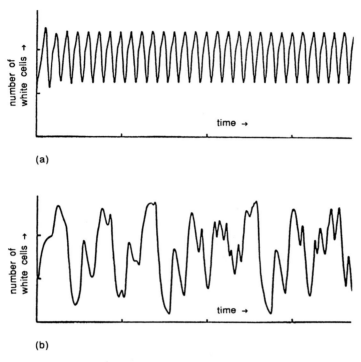

*Figure 4* Results from a model of the production of white blood cells: (a) short time delays give regular oscillations; (b) longer delays give apparently patternless chaos

The mathematical equations for this model are no more than twenty symbols long. They involve two adjustable numbers: one governs the response of the stem cells to the need to produce new white cells, one governs the time delay for maturation. It is not at all clear in advance what the solutions should look like or whether changes in these numbers can have any dramatic effects on the behaviour. The computer-produced figures, however, answer these questions without trouble. We might expect the white cell count to stabilise at some steady state – the correct number of cells. Instead, the computer calculations show several quite distinct kinds of oscillation in the cell count. For short time delays, the number of cells fluctuates periodically in a very simple pattern; for medium delays, the fluctuations remain periodic, but look much more complicated; for long delays, the oscillations show no obvious pattern at all. They appear neither periodic nor quasiperiodic. Indeed, they look random, irregular, patternless, and unpredictable.

This is a far cry from Laplace's triumphant determinism. Instead of having a perfect prediction of the entire universe, we have found unpredictable behaviour in an equation twenty symbols long describing, in a very simplified manner, the behaviour of one tiny subsystem of the human body.

What happens here is that the time delay effectively makes the stem cells respond to the wrong instructions. The white cell count alternately overshoots and undershoots the target value like a sticking thermostat: a big response too late rather than a small one at the right moment. It is not unlike Volterra's cyclic oscillations of fish populations – but there changing the numbers in the equations does not lead to chaos. Here, when the time delay is long enough, the system forgets its previous state, and by the time the instructions are acted on it has changed to an essentially independent state. In terms of our geometric picture, the feedback signal from the bloodstream turns over the number of white cells being produced, while the time delay stretches the dynamics. If there is not too much stretching, the turn-over acts like Figure 3b, producing a periodic response; but if there is enough stretching, we get something more like Figure 3c: chaotic mixing.

## THE BUTTERFLY EFFECT

I wrote above of mathematicians' dreams; one long-standing dream has been to predict the weather. In 1922 Lewis Fry Richardson dreamed up the Weather Factory – an army of young ladies housed in a vast building resembling the Albert Hall, operating mechanical desk calculators and communicating by semaphore and pneumatic tubes. Similarly, John von Neumann saw weather prediction as a major task for his new invention, the computer. By 1953 MANIAC at Princeton had confirmed that computers are capable of predicting the weather.

Getting the predictions right has turned out to be a little harder. One – and only one – of the reasons is a side-effect of chaos, which the experts call 'sensitivity to initial conditions' but whose nickname is the butterfly effect. It was discovered by the meteorologist Edward Lorenz in 1963. Lorenz was working on a system of equations that described a severely cut-down model of atmospheric convection. Unlike most scientists at that time, who were rather wary of computers, Lorenz had one of his own. Step by step it was calculating how his model behaved: a rather irregular series of oscillations, without much pattern to them.

He decided to extend the period of time covered by one of his calculations. Rather than starting at the beginning again, he read off the figures somewhere in the middle of the run, and started with those instead. The idea was to have some overlap, to check that the program was working properly. For a time the new computation agreed with the old, but then discrepancies began to appear. Soon the two calculations disagreed completely. At first he thought the program was faulty, but then the truth dawned. When he had entered the numbers for the second time, he had taken them from the computer print-out, which to save paper recorded only the first three decimal places. Internally the computer actually worked with six decimal places. Lorenz had assumed that the difference, an error of a fraction of one per cent, was inconsequential. But in fact, that tiny error had grown, slowly at first, but increasingly fast as the computer continued through ever more time steps.

Lorenz had a striking vision. Somewhere on Earth, a butterfly flaps its wings; a month later, the result is a hurricane. Of course, in the real world there are billions of butterflies. It is not clear whether this helps or makes the problem worse, and it is in any case hard to test whether real weather exhibits the butterfly effect. You have to run the weather twice, once with a wing-flap and once without, and see if it does something different. Yet the butterfly effect is definitely present within the computers that meteorologists use to forecast weather. The European Centre for Medium Range Weather Forecasting in Reading has carried out a programme of numerical experiments. Sometimes virtually identical initial data can lead, after only one week, to totally different forecasts.

Go back to our unfortunate duck in chaotic mode (Figure 3c). After only a few iterations, duck and non-duck are inextricably mixed. That is where the butterfly effect comes from: points close together get stretched apart and smeared until they lose contact and move independently. Would you like to predict whether the point at the centre of the square will be black or white after a hundred iterations? Would you bet money on your prediction? That is what weather forecasters are up against.

## STRANGE ATTRACTORS

Lorenz discovered something else about his equations. The traditional way to plot numerical data is as time series – a list of numbers, each a measurement of that quantity at a different instant of time, or a graphical representation of such a list. When plotted as a time series, the data produced by Lorenz's equations oscillate wildly. However, a lot more order emerges if the data are plotted not as time series but, following Poincaré's suggestion, as a phase portrait. Here this is a system of curves in three-dimensional space. For Lorenz's equations, these curves all lie on a relatively simple-looking geometrical surface something like the Lone Ranger's mask (Figure 5). No matter where you start the curve, as it follows the equations it homes in on the mask and then wraps round and round it.

Such a set is called an attractor, because all starting points are attracted towards it by the dynamics. The shrinking duck of Figure 3a, for instance, is visibly attracted to the fixed point. The attractor for a steady state is a single point, that for a periodic oscillation is a closed loop, and that for quasiperiodic motion is a torus – a combination of several independent loops. Lorenz had found what is now known as a strange (or chaotic) attractor – anything that is not a point, a loop, or a torus. The bestiary of strange attractors is vast but most of its pages are still blank. The commonest strange attractors at least are reasonably well understood, although sometimes the theorems lack complete

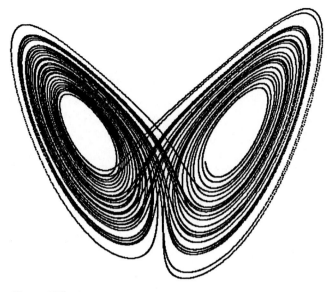

*Figure 5* The Lorenz attractor

proofs and instead are supported by computer experiments. The chaologist's bestiary has grown considerably in recent years. In consequence we can now recognise patterns in data that previously just looked random: for example, the wild oscillations of the white cell count above can be interpreted as motion on an attractor not unlike Lorenz's.

This ability to recognise new patterns in apparently random data is a double-edged sword. It suggests that many things we think of as being random may actually exhibit hidden order. No true scientist could pass up the challenge to dig that order out and see if it can be used for anything. However, the price we pay for this new knowledge is the realisation that very simple models can behave in very complicated ways. Finding the right equations may not tell us much about their solutions.

Does randomness really exist at all? We are back to Laplace. Philosophically, perhaps not: everything might be governed by fixed laws, which might be exquisitely simple or horribly messy. But operationally, yes, randomness exists whenever we cannot sort out the true pattern or decide if there is one. In nature apparently random behaviour is often the result of outside influences not built into our analysis. However, something similarly unpredictable can be generated internally by deterministic, but chaotic, motion. If the associated attractor is sufficiently complicated, sufficiently high-dimensional, the geometric patterns concealed within it may be beyond the visual abilities of the human brain.

## SIGNAL, NOISE OR BOTH?

The discovery of chaos has a profound effect on the manner in which experimentalists interpret their data. For example, as I have just explained, many experiments produce a time series of measurements of some physical quantity. Usually the time series is represented graphically as a moving trace: an electrocardiograph record of the heart, for example, is basically just a time series of electrical measurements.

All experimental measurements are subject to noise – unwanted variations in the signal caused by circumstances beyond the experimentalist's control (vibrations from outside sources, say, or fluctuations in temperature). In order to extract the essential behaviour from a time series, it is necessary to separate the signal from the noise. Until the discovery of chaos, this was in principle straightforward: the signal was the regular part, the noise was the random part. But now we have signals with no noise whatsoever that appear random. Depending on your picture of what should be counted as signal, and what as noise, you will be led to perform different mathematical manipulations, and obtain different conclusions. How can you tell which is right?

Classical methods for analysing time series begin by making assumptions about what constitutes a signal; typically they represent the signal as a combination of periodic motions. This method of analysis is not well adapted to chaotic signals because it begins by choosing particular patterns of motion which are actually somewhat alien to chaos. When chaos is present you can get signals that look just like noise. One person's noise may be another person's signal, and it all gets much more complicated. In consequence, more recent methods approach the problem of extracting the signal without committing themselves in advance to a list of standard patterns from which the signal must be composed. Instead they let the signal itself determine the basic patterns. The classical methods make the mathematician's standard mistake of forcing nature into a preconceived pattern; the new methods try to avoid it.

### TAYLOR-MADE

One of the most intriguing laboratory systems involved in testing theories of chaos is the Taylor-Couette system, a glass cylinder containing a cylindrical roller with fluid between the two surfaces; the roller is driven at various speeds and the flow pattern of the fluid is observed. The system was described by M. Maurice Couette in 1890, although several earlier investigators, including Isaac Newton and Arnulph Mallock, studied it. Its origins were practical: it was used to measure viscosity. Yet it has enormous intellectual appeal to experimentalists and theorists alike because it is such a simple idea with such puzzling consequences.

When the speed of rotation is low, the flow pattern is rather boring: the fluid just goes round and round slowly like the cylinder. However, in 1923 Geoffrey Ingram Taylor discovered that at slightly higher speeds, the fluid breaks up into layered vortices (Figure 6a). He devised a theory for this effect, compared it successfully with experiment, and thereby verified the standard equations of fluid dynamics. By the 1960s it became known that at still higher speeds a complicated sequence of changes takes place. First, the vortex boundaries become wavy; then the waves start to bob up and down; then the flow becomes turbulent, with very irregular waves all mixed up together.

There is a plausible overall description for all this: the rotating cylinder is a chaos generator, and the faster it goes, the less pattern there is. However, it is not that simple, for at higher speeds still,

*Figure 6* (a) Taylor vortices; (b) turbulent Taylor vortices

pattern again emerges. The flow once more becomes layered – but now the layers are turbulent. These turbulent Taylor vortices (Figure 6b), and the manner in which they appear, is quite a puzzle: order and chaos in the same object.

The chaotic aspects of turbulent Taylor vortices have been studied in a beautiful series of experiments by Tom Mullin at Oxford, using the methods mentioned above to recreate the geometry of the underlying attractor. The dramatic and elegant results are shown in Figure 7. The geometry is that of a large sphere with a narrow tube bored through the middle. The system begins by spiralling down the tube, and then returns over the surface of the sphere in huge swirls to re-enter the tube again. However, the position of successive sweeps varies pretty much randomly. This particular attractor is known to

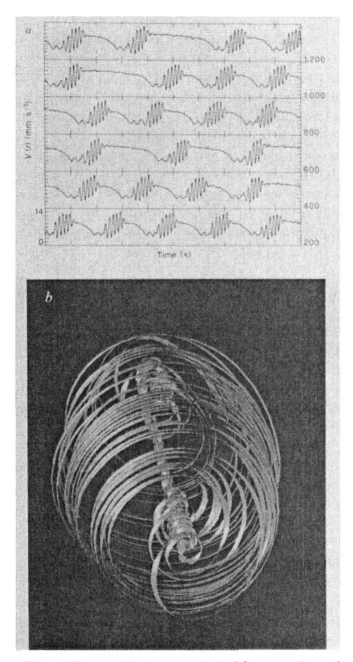

*Figure 7* Strange attractor reconstructed from experimental data in turbulent Taylor vortices

mathematicians so the experiment pins down the mathematical mechanism that is involved in this type of turbulence: a now-familiar item from the chaologist's new bestiary.

Traditionally, order and chaos are two distinct polarities. This black and white picture of the universe around us is now seen to be misleading: there are shades of grey, a continuous spectrum of behaviour ranging from total order to total chaos. However, even this picture retains traces of the black/white distinction. The observed state is either ordered or chaotic, but we do not expect it to be both! Turbulent Taylor vortices show that our understanding of the relation between order and chaos must be expanded still further: a system can exist in a single state that simultaneously displays aspects both of order and of chaos.

## FUZZY SYMMETRY

What possible mechanism could produce such a curious mix? What we currently have is not so much a final answer as a very plausible suggestion, a fundamental mathematical mechanism that has all the right features. It is not yet certain that it is responsible for the patterned turbulent states in Taylor-Couette flow – the calculations are too hard! – but it definitely plays a similar role in other systems, notably electronic circuits.

The key to patterned turbulence appears to be its relation to symmetry. The Taylor-Couette apparatus is highly symmetric, and so are its flow patterns, even the turbulent ones. The most obvious symmetry is rotational: if the entire apparatus is turned through some angle, then to all intents and purposes it looks exactly the same. It also has reflectional symmetry in a horizontal plane through its middle. Finally, it has a useful approximate symmetry: a very long cylinder can be translated along its axis. This is not an exact symmetry because one end pokes out; but that does not affect what happens in the middle by very much. In fact the standard mathematical model of the Taylor-Couette system employs an infinitely long cylinder for which translations along the axis are exact symmetries.

What about the flow-patterns? It turns out that each has its own set of symmetries. Couette flow has precisely the same symmetries as the apparatus. Because it is patternless, it looks exactly the same if it is reflected, rotated or translated. Couette flow has so much pattern (it is the same everywhere) that effectively it has none. Taylor vortices look exactly the same if they are rotated: each individual vortex has circular symmetry. The same holds for the reflectional symmetry in a horizontal plane: if you turn the stack of vortices upside-down, it looks the same as before. Because of the striped pattern, the only translational symmetries of Taylor vortices are those that move the cylinder through a whole number of stripes.

What about the turbulent flow? Technically speaking, these have no symmetry. However, if you ignore the fine structure of the turbulence and treat it just as additional texture, then the flow has a great deal of symmetry. Turbulent Taylor vortices have visible structure, both vortex layers and stripes. Apart from turbulent fine detail, the flow is just fuzzy Taylor vortices and – forgetting the precise details of the fuzz – it has the same symmetry as Taylor vortices. Mathematically there is a problem: how do we filter out the fine texture? There is no well-developed theory of fuzzy symmetry: the answer is to reinterpret everything so that the fuzzy symmetries become genuine ones. To do this we go back to basics and ask a simple but important question. How does the symmetry of a system affect its dynamics? In particular, what effect does symmetry have on chaos?

Here we must decide what it means for a dynamical system to be symmetric. The answer is that at each instant of time symmetrically related points must always move to symmetrically related places. A consequence of symmetric dynamics is that the corresponding attractor, or attractors, may also have symmetry. This may incidentally be less than that of the entire system, a phenomenon known as symmetry-breaking. Dynamics with symmetry has been explored by many mathematicians: in particular, symmetric chaos has been studied in computer simulation by Pascal Chossat, Mike Field, and Martin Golubitsky. They have concocted discrete dynamical systems having attractors whose symmetries are the same as those of a regular

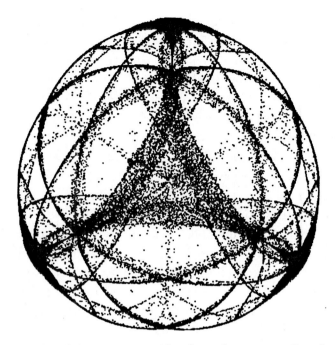

*Figure 8* Chaotic attractor with polygonal symmetry. Though black and white here, stunning coloured versions of this figure have also been produced

polygon (Figure 8). Moreover, symmetrically related attractors can sometimes collide to create a new attractor with more symmetry. This is very similar to the creation of turbulent Taylor vortices and it is conjectured that the basic mathematical mechanism is identical. In the world of ordinary attractors, symmetries tend to be lost as attractors break up; but in the world of strange attractors, symmetries tend to be gained as attractors merge.

This simple example reveals two fundamental principles: first, that symmetric dynamical systems often have several distinct attractors, related by symmetry transformations; and, second, that these individual attractors can merge to create a strange attractor with more symmetry.

So now we understand how symmetry and dynamics together can

create attractors that combine aspects of order (symmetry) and disorder (chaos). At Warwick University's Nonlinear Systems Laboratory, Peter Ashwin and Greg King have performed experiments to confirm these two principles. Their apparatus is an electronic circuit in which three or more identical oscillators are coupled together symmetrically and observed with an oscilloscope. The resulting data are processed by computer to exhibit the underlying attractor in a way that brings out the symmetries. They have found chaotic attractors with various symmetries, and symmetry-increasing collisions.

## CHAOTIC QUILTS

Let me end with a personal story. I spent March 1990 in the United States, much of it at the Institute for Mathematics and its Applications in Minneapolis. One morning Martin Golubitsky and I were eating breakfast in the hotel and were regaled with a human interest story about how textile patterns can be made using computers. We had no choice but to pay attention, because continental breakfast – twenty varieties of doughnut and nothing else – was included in the cost of the room. To aid digestion the management had thoughtfully provided a huge television screen with a powerful sound system smack in the middle of the main wall. We already knew that the Chossat–Golubitsky–Field method created excellent icons: Martin had used one as a T-shirt design for the *Dynamics Days Texas* conference. But textile patterns are usually repetitive – quilts rather than icons. As the programme resonated through our skulls we realised that we could also design quilts using chaos, finding a dynamical system with the symmetry of a tile pattern and persuading it to go chaotic. Such a process always leads to a pattern that tiles the plane, repeating periodically and matching perfectly at the joins.

In the peace and quiet of Martin's hotel room we decided to enhance the symmetry characteristics of this pattern by demanding that on each tile, the dynamics have square symmetry. We worked out a formula for the simplest dynamical systems that meet all of these requirements. It involves five arbitrary numbers, one of which must be an integer. To find interesting quilts, you have to choose the right numbers. Which? We had no idea. Chaos is like that.

So, using a lap-top computer, we typed in the formula, guessed plausible values of the numbers, hoping to find a few chaotic sets with square symmetry – and to our surprise discovered one straightaway. The reason for this early success rapidly became clear: nearly every set of numbers that we tried worked. Some – ironically, given our hi-tech approach – look like floral patterns; some strongly resemble batik; and some (Figure 9) look like a stained glass window influenced by both Christian and Islamic images. All of these varied

*Figure 9* Chaotic quilt design by Mike Field and Martin Golubitsky

designs were produced by the identical, simple procedure, merely by adjusting a few numbers.

The results have two important implications over and above their aesthetic appeal. The first is that symmetry and chaos – pattern and disorder – can coexist naturally within the same simple mathematical framework. The second is that the chaotic patterns produced by this technique look complicated, yet they are prescribed by a short computer program and a few numbers. Their information content is thus actually very small. Normally we think that it takes a lot of information to specify a complicated structure – but here it does not. There are two possibilities. One is that complexity is not conserved: it can be created from nothing. The other is that forms that appear to be complex may not actually be so. The latter is, to me, a more appealing interpretation: the quantity of information needed to *de*scribe an object may be more than that needed to *pre*scribe it. So a by-product of our messing about with quilts is a serious question about the nature of complexity.

## FURTHER READING

Field, Mike, and Golubitsky, Martin, 'Symmetric chaos', *Computers in Physics* 4(5) 1990, 470–8.

Gleick, James, *Chaos: Making a new science*, Harmondsworth: Penguin Books, 1989.

King, Gregory, and Stewart, Ian, 'Symmetric chaos', in *Nonlinear equations in the applied sciences* (eds. W. F. Ames and C. F. Rogers), New York: Academic Press, 1992, 257–315.

Stewart, Ian, *Does God play dice? The mathematics of chaos*, Oxford: Blackwell, 1989.

Stewart, Ian, and Golubitsky, Martin, *Fearful symmetry: is God a geometer?* Oxford: Blackwell, 1992.

# 3

## Comets and the world's end

*SIMON SCHAFFER*

In the year 838 a great comet appeared over western Europe. At the Emperor's court at Aachen, the phenomenon provoked considerable debate. The Emperor consulted his astrologer. Drawing on the best contemporary learning, the astrologer knew that the comet signified the monarch's death and turmoil in the state. He decided to keep silent. The Emperor saw through this ruse. He concluded, piously, that the comet was a warning from God. The court was ordered to its devotions, alms were distributed, and the Emperor then went hunting in the Ardennes. The astrologer summarised: 'more game than usual are said to have been killed, and all the projects the King undertook at that time turned out successfully'. This was over-optimistic: within two years the Emperor was dead and his realm divided between his warring heirs.

The episode is characteristic. Comets have always been ambiguous signs, prompting extraordinary human efforts to make them meaningful. A century ago, the distinguished British astronomer John Herschel (1792–1871) lectured on just this theme: 'there are so many things in the history of comets unexplained, and so many wild and extravagant notions in consequence floating about in the minds of even well-informed persons, that the whole subject has rather, in the public mind, that kind of dreamy indefinite interest that attaches to signs and wonders than any distinct, positive, practical bearing'.

In good Victorian fashion, Herschel was much taken with the possible practical use of these bodies. He tried to find evidence of divine design, or mundane advantage, but he drew the worst possible conclusion: 'hitherto, no one has been able to assign any single point in which we should be a bit better or worse off, materially speaking, if there were no such thing as a comet'. This was a gloomy summary of centuries of earnest work: what had been the most significant opportunity for successful prediction was now dismissed as utterly trivial.

## PROPHECY: A QUESTION OF TRUST

My story is concerned with a pair of changes in the way we predict. First, we see a change from the use of cometary appearances as a chance for soothsayers to predict the future course of politics, the weather and other features of immediate interest, to cometary transits as themselves foreseeable threats to this planet. Comets had been occasions for prediction; then they became predictable. Second, there were complicated changes in the kind of person who could be trusted to make predictions. Court astrologers and popular almanac makers gave way to expert advisers brandishing the tools of statistics and physical science. So comets provide us with a good way of seeing how prediction depends on politics. Comets were publicly available signs, representations of heavenly wrath. Somehow or other Europeans learnt to stop worrying and trust the astronomers. In this process, the great divide between traditional and modern culture was created.

Yet things are not in practice so simple. As Ian Kennedy emphasises in Chapter 5, prediction is intimately bound up with relations of trust and jurisdiction. To contemplate the business of prediction in isolation from its target audience is to miss the force of these problems of credit and authority. In the realm of prognosis and therapy, Kennedy shows us, the contest between parliament, law court and medical consulting room illustrates this fundamental moral and social principle. My purpose is to draw attention away from the abstract question of the reliability of prediction and towards the everyday problem of the trust we invest in our favourite predictors.

Each culture privileges a select group of soothsayers, and the choice the culture makes tells us much of its values and interests. In our own generation, we have witnessed the rapid succession of town planners and econocrats, cost–benefit analysts and political ecologists, military analysts 'thinking the unthinkable' and so-called 'chartists' in the City, all reading entrails and deploying allegedly credible futurological machines.

Experts need their own private world where ignorant outsiders cannot penetrate. The very obscurity of the sums in which cost–benefit analysts or astrologers engage helps give them impressive authority. But the expert predictors also need outsiders' trust: they need to show that the terms they use are, in some way, connected with what matters to their customers. Cost–benefit analysts evaluating the construction of a new airport invent what they call a 'weighting for posterity', even though it is always hard to guess what value posterity would place on the wildlife, or the old buildings, destroyed by the process of airport construction. Astrologers often try the same technique, making the most commonplace concerns feature in their complex charts. So predictors have to move between specialist technical work and public, widely accessible, concerns. One mistake is to suppose that the culture of the wider public has no effect on the specialist predictors; it does. A lesson of the comet stories is that the most apparently technical estimates of cometary science are very sensitive indeed to public needs and attitudes.

## 'TO DIFFUSE IN THE PUBLIC MIND SOUND AND PRECISE NOTIONS'

Expert discussion about cometary fears and forebodings never ceased to emphasise the importance both of specialist private techniques of calculation, and everyday public concerns about the world. Victorian astronomers, John Herschel's colleagues, understood that the construction of a secure space for cometography required massive transformation in public opinion. Thus in his influential account of comets and their prospects, the French physicist and radical politi-

*Figure 1* A comet collides with the Earth: French print, 1857

cian Dominique Arago (1786–1853) claimed that 'comets no longer frighten the public – this is a result respecting which science has certainly the right to congratulate itself'. But as a protagonist in the revolution of 1848 he was well aware of the uneven career of what he counted as reason: 'to diffuse in the public mind sound and precise notions will be the best means of preventing writers without authority from abusing it when one of those mysterious bodies appears suddenly in the heavens'.

Arago's strategy was clear. Recognising the immense public interest in cometary transits, he wished to reform public opinion principally in order to allow the people to discriminate between reliable experts and charlatans. Precise prediction of cometary paths would disqualify experts' rivals. This explains why, in a work principally devoted to the transit of Comet Halley in 1835–6, a transit which caused great public concern, Arago included answers to such questions as whether a comet could collide with the Earth, or orbit round it, or affect its weather. It was important for Arago to discuss meteorology, since many Frenchmen spent large sums on the vintage wine of 1811, the so-called 'comet wine', grown following the excessive heat of that summer. Arago stated that there was no evidence that comets affected the vineyards: 'I shall now leave it to the reader to judge', Arago wrote, 'if the wine-growers would ever be justified in founding any hopes on the apparition of a comet!' But this joke concealed a sterner task. Arago and Herschel both strove to dispel crucial and widespread claims that comets were threats to our security, that they would presage or cause the world's end.

As we shall soon see, early modern Europe viewed comets with awe and terror, not simply under the traditional rubric of heavenly warnings, but as real physical threats to the Earth. So our story concerns itself with an irony, which I wish to underline. Once comets began to cease being purely astrological symbols, they rapidly acquired the status of palpable threats. The public held that physical bodies could wreak a degree of havoc that celestial ciphers could not. So Arago made comets physical (hence not mere signs), but benevolent (hence not real threats). He discussed the question of cometary habitability, concluding that humans had indeed survived the extremes of temperature and pressure which they would have to undergo aboard a comet. In order to illustrate both the heroism of scientists and the conditions humans could tolerate, he cited Halley's trip in a diving bell, Gay-Lussac's ascent in a balloon, and John Franklin's appalling experiences in the Arctic. 'I do not profess to deduce from these considerations the conclusion that comets are peopled by beings of our species. I have only presented them here in order to render their habitability less problematical.'

Arago was not concerned to disabuse the French of their bizarre ideas about comets' residents. He was concerned to make sure that they knew that scientists, and no one else, could give them details of these residents' life style, and to suggest that life, rather than death, was a comet's accompaniment.

The work of Arago and Herschel defines one end of our story. By

56

*Figure 2* Honoré Daumier, *A surprise*: lithograph, 1853

the mid-nineteenth century, expert predictors were seeking to connect popular culture with their own secure knowledge, and thence to secure both their status and that of the world itself. Cometography would be safe from charlatans, and the world would be safe from comets. We can use this picture of predictions to interpret the way that work on comets changed from divination of the meaning of signs to discussions of the world's end, and thence to the relaxed securities of Victorian astronomy. A complete exposition of this process would surely require a comparably detailed examination of the ways in which European culture switched its pattern of credit from court society and church establishments to that of the expert professions. We should also have to document the process through which patrician and plebeian cultures split, creating new notions of superstition and truth.

Two great events have often been seen as marking this transformation in cometary science. First, in 1577, Tycho Brahe (1546–1601), the Danish nobleman and astronomer, decided that the great comet of that year must be further from us than the moon. Hence, he concluded, the Aristotelian notion that comets were like meteors, fiery exhalations just above the Earth, was false. For a century and a half after, the character of comets remained in question. Second, in 1759 astronomers successfully predicted the return of Comet Halley, thus vindicating the truth of Newton's claim that comets were predictable and orderly members of the solar system, moving in very eccentric ellipses around the Sun.

This story has a pleasing symmetry, but it simply fails to match the manifest changes in cometary significance. Tycho's work did not mark the end of astrological cometography; and the prediction of Comet Halley's return did not terminate fears of these errant bodies. Instead, different cultural forces also need to be considered in order to understand how the position and predictions of astronomers and their rivals developed.

## 'THE ETERNAL SABBATH IS AT HAND'

Tycho's work on the great 1577 comet was indeed a benchmark in the history of cometary prediction. But it is not best understood as the dawn of sober rationality. The great comet was as bright as Venus and possessed a tail of 22 degrees in length. Such appearances were indispensable resources in Renaissance court culture. They were wonders to be presented by experts to their patrons, provided this was done with suitable care and cunning. It is unsurprising that the astute and ambitious Tycho seized on this new sign. Tycho's literary and learned treatise of 1572 on the new star of that year must have attracted the patronage of his King, Frederick II. Between 1576 and 1579 he received a very impressive range of royal endowments, including the island of Hven, where his observatory, Uraniborg, would be built. In return, Tycho produced horoscopes for each royal son, and an annual prognostication for the court. The great comet appeared over Denmark at

Martinmas, 11 November 1577. The King was supplied with a pamphlet by the Copenhagen theologian and Lutheran extremist Jorgen Dybvad, Tycho's ideological rival.

The importance of astrological advice for Danish politics cannot be overestimated: Anne of Denmark, Frederick's daughter and wife of James I, had important occultist connections, while Christian IV, her brother, built an enormous magical palace at Frederiksborg. Hence the utility of the great new comet. In a German pamphlet of 1578 and a massive Latin treatise of 1588, Tycho outlined its position and character, and predicted its consequences. He noted the persistent interest humans had displayed in great comets. He endorsed the views of the notorious German chemist and wonder-worker, Paracelsus. The heavens were formed of a subtle and translucent element. Aristotle was wrong to suppose that this could not breed change. Comets were 'pseudoplanets', supernatural monsters specially created by God, sent as signs to warn humanity. But like planets, their motions and characters were a proper topic for astronomy.

In a magisterial survey, Tycho synthesised all available observations of the comet. He subjected all his contemporary astronomers to careful professional scrutiny. After a lengthy sequence of critical calculation based on the thirty nights when the comet was visible, carefully sifted for maximum polemical effect, Tycho concluded that the comet must be five times higher than the moon, and, in a radical set of cosmological claims, urged that it moved round the sun in the opposite direction to the inferior planets. His predictions for its consequences were modestly based on 'the ancient experienced astrological authors, and this can be done without superstition or mummery'. His enemy Dybvad was castigated as a 'pseudoprophet', and Tycho tailored his predictions against those of the orthodox Lutherans. This sign would govern the world order until the next great astrological event, the conjunction of Saturn and Jupiter in Aries of 1603. This maximum conjunction, which took place roughly once every eight centuries, had therefore happened only six times before in the history of the world: 'The eternal Sabbath of all Creation is at hand in this seventh greatest conjunction.'

*Figure 3* Jiri Daschitzsky, broadside of the comet of 1577 seen at
Prague: woodcut. This was the comet observed by Tycho

While conceding the unpredictability of the world's end, Tycho
achieved at least two key goals with this comet: first, he defined the
skills required for membership of the select community of disciplined
astronomers and scotched his immediate rivals; second, he used it to
presage a need for the reformation of the world system on the basis of
what he called 'manifest observations with proper instruments', as
opposed to scholastic 'subtle rational arguments'. Tycho deployed
highly sophisticated mathematical techniques to give his community
a new role in the world order.

Tycho's programme enabled a complex compromise in which court
astronomers and their colleagues figured out ways of predicting
events on the basis of their mastery of the causes operating in heaven.
Thus Johannes Kepler (1571–1630), a master of causal astronomy,
used his Lutheranism as a means to understand comets as signs of
God's grace, and saw them not as heralds of specific events on Earth
but as warnings of the end of this world and harbingers of the world to

come. In his *More certain fundamentals of astrology* (1601), Kepler combined his almanac for the coming year with a learned discourse on the bases of this science. 'Astrology clearly has some say in political and military matters', he concluded. He even had a story which explained why astrology worked: humans' minds were made of a stuff similar to that of the heavens. 'The mind of its own accord shares feeling with the sky, because it possesses cognition with light and harmony. Furthermore, since a person is a social animal, dispositions are particularly oriented to a public undertaking when the rays of planets are oriented geometrically in the heavens.'

Nor did Kepler miss the chance to define the appropriate deference he could expect from his patrons: 'this task may be undertaken more accurately if there are ready at hand the horoscopes of those who (if I may use a phrase of Tycho) govern public destiny'. Hence arose, we may conclude, the characteristic discourse of predictive cometography of the Baroque world.

Experts were required to balance the demands of their patrons and of their culture in reading the heavens. For those in search of a clear-cut division between rational astronomy and occult astrology, this may be a disappointing message. Magical resources remained important. For example, the Cambridge divine John Edwards used his tract, *Cometomantia* (1684), to condemn the absurdities of judicial astrology and the heresies of almanac makers. But he also teased out the causal chain, the property of expert augurers, which allowed the heavens to determine earthly affairs: he agreed that comets would dry the atmosphere and produce 'dearth, scarcity and famine. And, as the inevitable effects of both, we must expect sickness, diseases, mortality, and more especially the sudden death of many Great Ones', because the wealthy were easily struck down after the indulgences in luxury to which they were prone. Hence this weakness of the powerful was just what made them listen to the heavens and vulnerable to the heavens' effects. These factors, in their turn, would guarantee attentive audiences for the astronomers and would discriminate their concerns from those of the vulgar.

## 'WE MAY KNOW WHETHER
## THE SAME COMET RETURNS'

Edwards's Cambridge and puritan milieu of the 1680s brings us immediately to the setting in which Isaac Newton (1642–1727), the university's mathematics professor, inaugurated his own work on what had now become the central problems of predictive celestial mechanics. We must consider the sense of Newtonian cometography in the light of the credibility of experts' predictive powers. Throughout Newton's lifetime, comets were a central concern for a wide range of authorities: priests, astrologers, journalists, astronomers and natural philosophers. The frequent transits of major comets during that period repeatedly prompted explosive pamphlet wars on the topic of cometography. Two issues mattered: first, were comets permanent or transient occupants of the skies? Second, what was their significance for the inhabitants of the Earth? Both were questions whose sense changed radically in the seventeenth century. The comets of 1664 and 1665 were often blamed for the plague and fire of the following years.

The sources Newton read when spurred to initiate his astronomical researches after the cometary transit of 1664 agreed that permanent objects must move in closed orbits, while transients would move in open or rectilinear paths. By 1679, Newton had convinced himself of the validity of this conventional wisdom. In the next five years he was presented with fresh opportunities to contemplate cometography and to rethink this view. On the one hand, correspondents engaged him in debate about the system of the world, stimulated at least partly by the comets of 1677, 1680 and 1682. On the other, the political and theological crisis which wracked Britain during just these years elicited streams of texts on political astrology and polemical philosophy which made use of the signs of heaven and re-theorised the heavenly mandate of the monarchy.

John Flamsteed (1646–1719), as the new Astronomer Royal, worked hard at Greenwich to transform astrological uses of comets by arguing that they were periodic. He guessed that the comet of 1677 would

*Figure 4* Broadside of the comet of 1680 fulfilling a prophecy of
Nostrodamus concerning the fall of Rome: German print

return in 1689. He also responded actively to the furore about the
comet of 1680 which coincided with the political crisis of the Popish
Plot and the envisaged succession as monarch of James, Duke of York,
an avowed Catholic. The young Edmond Halley (1656–1742), then
travelling in France, also commented on this astrological and religious
connection. One comet had been visible in November 1680 while
another appeared from behind the Sun in December. A question
strenuously argued between Halley, Flamsteed and Newton the following spring was whether these comets were one and the same
object. Flamsteed, keen to support the natural philosophical claim
that comets returned, insisted they were but one body. Newton
denied this and, following an orthodox view, held that comets were
transient, so could not incline at the Sun through as great an angle as
Flamsteed's hypothesis would imply.

In 1679 to 1681, therefore, Newton did not hold that comets return,

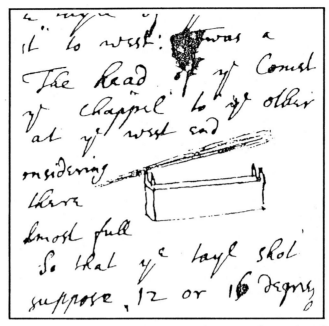

*Figure 5* Newton's representation of Babington's observation of the comet of 1680–1 over King's College, Cambridge

nor did he claim that they were moved by an attractive force situated in the Sun. This suggests that Newton's cosmology was reconstructed, particularly in respect to cometography, after 1681 and before the heroic period of drafting for the *Principia mathematica* initiated with Halley's celebrated visit to Cambridge in the summer of 1684. Only then did Newton decide that 'we may know whether the same comet returns time and again'. Flamsteed ironised about Newton's belated but decisive concession that the comet of 1680, for example, was but one object: 'he would not grant it before see his letter of 1681'. From this moment, in the winter of 1684–5, the puzzle of computing the cometary orbits, which Newton already assumed were elliptical, became one of his principal preoccupations. His assumption that they moved in ellipses was reached with little empirical warrant. Ultimately, during 1686, when composing the final book of the *Principia*, he found an incredibly accurate graphical method, but his cosmological views about comets were already in place in the first version of this book, a *System of the World*, probably written in autumn 1685. There, as in correspondence with Flamsteed and Halley of this time, he argued that comets all move in ellipses for which parabolas may serve as good approximations. This was specifically applied to the comet of 1680–1, which he held truly moved in an ellipse and would therefore return.

The comet of 1680–1 became almost as celebrated a Newtonian prediction as Comet Halley itself. Thus Edward Gibbon characteristically paused during his *Decline and Fall of the Roman Empire* (1776–88) to guess that at the comet's return 'in the year 2255 Newton's calculations may perhaps be verified by the astronomers of some future capital in the Siberian or American wilderness'. From the 1690s, Halley and Newton also worked on further cometary data, developing what may be called an historic method for estimating cometary periods – comets with similarly oriented and shaped parabolic approximations, moving in the same direction, could be identified with each other. If such comets recurred at regular intervals, a periodic body had been located and its period could be found. This period would then allow a more exact, elliptical shape to be computed.

This was the burden of Halley's work which generated the inexact prediction of the return of the comet of 1682 for the late 1750s. The goal of this work was to reinforce the implication that all comets returned and that returns could be made susceptible to analysis by skilled Newtonian cometographers. Ultimately, these claims received influential publicity in Halley's celebrated *Synopsis of the astronomy of comets* completed in March 1705. The tables Halley printed there were used in successive editions of the *Principia mathematica* and reprinted in many astronomy texts. But we must not be misled by this publicity: the methods were difficult and by no means commanded immediate assent. The great French astronomers Lacaille and Delambre both held that Halley's exposition of the historic method was so 'unintelligible' that few astronomers could actually follow it in the eighteenth century.

We can start to see that Newtonian cometography did not mark the end of cometary significance. It is important to stress the extraordinary political role which Newton gave to comets. From the mid-1680s, he began to argue that true old Babylonian cosmology had held that the Sun was in the middle of the world and that space was empty – Newtonian, in fact. The Chaldeans and the earliest Greeks had believed that planets and comets orbited the Sun under gravitational influence. Pythagoras had expressed these truths in his mystical imagery of the harmony of the spheres. But then 'this philosophy was discontinued – it was not propagated to us and gave way to the vulgar opinion of solid spheres'. Newton reckoned that this explained why Aristotle and his successors wrongly reckoned that comets moved between the Earth and the Moon. Once solid spheres had been introduced and the Earth placed at the world's centre, a sinister political conspiracy emerged. The planets were now deemed to move through the action of dead men's souls, especially the souls of dead kings, who were thence glorified through divine right: 'so ready was the ambition of Princes to introduce their predecessors into the divine worship of the People to secure to themselves greater veneration from their subjects as descended from Gods'. This meant that scholasticism, Papism and the divine right of kings were all based on bad cometography. Restoring proper astronomy would help restore proper politics and proper religion.

These arguments had particular resonance, of course, in a period of contest between Cambridge and King James, in which Newton took a very active part on the university's side. Newton communicated his opinions about the link between anti-Catholic politics and true cometography to most of his closest disciples, and his views were propagated via them to many early eighteenth-century readers. Similarly, he gave comets renewed cosmological roles in the preservation, restoration and increase of activity in matter distributed in space. In the 1690s, Newton said that comets acted as agents in the 'Growth of new Systems out of old ones'. At exactly the same period, in 1693–5, Newton discussed with Halley views which the latter had begun to develop in the late 1680s on the possible cometary cause of the Deluge. In subsequent editions of the *Principia* Newton incorporated suggestions that comets might increase the mass of the Earth and so cause the gradual acceleration of the Moon. Already, in 1687, he published his view, comparable with that of his colleague John Edwards, that comets helped restore life on Earth: 'comets seem to be required, that, from their exhalations and vapours condensed, the wastes of the planetary fluids spent upon vegetation and putrefaction and converted into dry earth, may be continually supplied and made up'.

Comets had now become the principal transmitters of life and restorers of vitality in the heavens and on Earth. In the third edition of the *Principia*, following conversations with Halley, Newton also rewrote his view that comets would fall into stars. In 1725–6 he announced that such events would produce novae, and it was in this context that, according to his nephew John Conduitt, Newton joked about the incendiary effects of such events on the Earth. The full role which had emerged for comets, substantiated by Halley's data and with Newton's cosmological scheme, then provided eighteenth-century commentators with much fruitful opportunity for predictive and pessimistic discourses on cometary fate.

In his astronomy textbook of 1702, the Oxford professor David Gregory (1659–1708) insisted that 'those things which have been observ'd by all Nations, and in all Ages, to follow the Apparition of Comets, may happen; and it is a thing unworthy of a Philosopher to look upon them as false and ridiculous'. Citing Gregory's view, both the French philosophers Maupertuis and Buffon held that the Newtonian cometography rightly restored comets in their full terror; in 1761 the Alsatian natural philosopher Johann Lambert (1728–77) reckoned that enlightened cometographers were turning themselves into what he called 'authorised prophets'. Lambert's striking phrase was well-chosen. Prophecy and cometography were intimately connected in Newton's scheme. Just as he made them return, so he made them objects of enormous significance in the history of life in the cosmos and on Earth. Comets had been used by radicals, sectaries and the godly as marks of the political and theological crisis enveloping the state. Newton held that with a restored cometography and cosmology, the philosophy which underpinned that corrupted state and church would be discredited.

I have made much of the political and apocalyptic significance

*Figure 6* God directs a comet towards the Sun, producing the planets: from the Comte de Buffon's *Histoire naturelle*, Paris 1749

which Newton gave his cometary cosmology, and I have emphasised the ways in which his account of comets fitted them into a world-system which might presage the end of all things. Enlightenment interpreters of Newtonian cometography put his work to their own uses. They did not share his commitment to eschatology. But they did agree that comets played a fundamental role in the cosmos. And they did agree that this role might well be destructive. The recovery of Comet Halley was certainly treated as a triumph for Newton's world view. But there was still no obvious division between secure

cometography, derived from astronomical science, and the speculations on scriptural interpretation and imminent catastrophe which
drew much public attention throughout this period.

### 'OF COMETS WHICH
### MAY CRASH INTO THE EARTH'

We have now charted the resources which were used to make Newtonian cometography a reputable science. Its apostle was the doyen
of French academic physics, Pierre Simon de Laplace (1749–1827).
Laplace understood predictability as the chief goal of celestial
mechanics. He made an explicit connection with the custodial role of
the expert. The confidence which Newtonian astronomy gave the
savant was to be contrasted with the errors of the vulgar, those whose
superstition prevented them from understanding the long causal
chains which governed all natural events. This is how he made his
case:

> An intelligence which, for a given instant, knew all the forces by
> which nature is animated, and the respective situation of the
> beings which made it up, if furthermore it was vast enough to
> submit these data to analysis, would then embrace in the same
> formula the movements of the greatest bodies of the universe
> and those of the lightest atom: nothing would be uncertain for
> it, and the future, as the past, would be present to its eyes. In the
> perfection which it has known how to give to astronomy, the
> human mind offers a feeble sketch of this intelligence ... All its
> efforts in the search for truth tend to bring the human mind
> ceaselessly closer to the intelligence which we have just
> imagined, but from which it will always remain infinitely
> removed. This tendency, proper to the human race, is what
> makes it superior to the animals and its advances of this kind
> distinguish nations and centuries, and make up their true glory.

This classic formulation of determinism, taken from the published
version of Laplace's celebrated lectures delivered in Paris in the
1790s, has been the object of a long tradition of commentary on the
overweening ambitions of science. However, we must read this

passage with care and in context. Laplace presents his audience with a version of the principle of sufficient reason. All events have rationally describable causes. But he then notes that not all recognise the truth of this principle. Such rationality is the prerogative of the elite. The vulgar are just those who lack this understanding.

These claims by the savants of the high Enlightenment provide the right context in which to place the story of prediction and confidence. It is therefore worth while to describe the expert prophet's place just before the French Revolution. Measurement of belief and measurement of nature fitted together in the enterprise of the late eighteenth-century savants. Peter Burke has argued that it was at this period that the very term 'superstition' begins to change its sense, from a dangerous falsehood, worthy of contest, to a silly popular belief, worthy of satire, or, at best, folkloric study. Cometary astrology changed in just this way, and became an object of elite scrutiny rather than hostility. What was real in this popular culture was simply disqualified by the savants. The same applied to 'falling stones', a popular error ridiculed in the Paris Academy of Sciences until after an academic report of 1794: when 300 rural witnesses signed an affidavit supporting the existence of such meteorites the physicist Bertholon commented on 'popular sensations which can only excite the pity, not only of physicists, but of all reasonable people'. The cases of meteorites clarify the fact that social status dominated the assessment of different predictions and different beliefs. This mattered especially in late eighteenth-century France because of the political meanings of fear and terror, ultimately during the great fear of summer 1789 and the massacres of September 1792. The experts' job was obviously to categorise elite terror and popular delusion as irrational, the product of imagination, and contrast this with the real.

Interest in comets provides a fine example of the Laplacian dream and its politics. Astronomers frequently told their audiences that proficient observers of the heavens saw things differently from common men. Both observation and calculation gave astronomy its status as the highest discipline. To preserve this status it had to challenge public fear and also to exploit it. Thus the astronomer Jérôme

Lalande's *Reflections on comets which can approach the Earth* was published in 1773 amidst enormous public interest and terror. Lalande had been a protagonist in the calculations of the return of Comet Halley in 1759, and his views were taken very seriously.

The news that the authoritative academician Lalande had predicted an imminent cometary collision spread throughout Paris and the provinces. There were stillbirths in Normandy, an archiepiscopal mass in Notre Dame, and Lalande was arrested for a breach of the peace. He commented that 'these popular clamours reached the point of terror and I believed I owed the public an explanation capable of reassuring it'. The Academy itself played down the strife: Lalande's memoir was 'merely hypothetical, and based on possibilities, but of one to 64,000, so it could not disavow the recognised principles of Astronomy' which, so it was reported, 'produced an even worse effect, in confirming what he had put forward'. The 'terror' he initiated highlighted the instability of stabilist celestial mechanics, and made the term 'astronomical revolution' more than ambiguous. The political message was not missed. Both the public journals and the Académie's own *Mémoires* also satirised the credulity of its own audience: 'the heads of our lady friends are uplifted and we have had a lot of trouble in calming their terrified imaginations'. An influential academician wrote that the effect of the memoir highlighted the problem of explaining the meaning of the word 'probable' to the public: 'in using the word *impossible* in the sense of common language, we can firmly say that the encounter of a comet with the Earth is impossible and that we have nothing to fear from these stars'. The academicians continued that 'the people whose fear was the liveliest were the first to stop fearing, because they were the first to forget that comets existed; since it is a gift of Nature that the weakest imaginations are also the most fickle'.

A detailed analysis of the probabilities involved in Lalande's supposed scenario was presented to a committee of the Académie, including Laplace, in August 1774. The committee was sympathetic: 'the ignorant and timid vulgar, having no other reason to assure themselves against natural phenomena which are rather singular, except the example and the authority of enlightened people, become alarmed very easily'.

Lalande's memoir was disturbing not for his analysis of the relation between cometography and celestial mechanics but because he had inadvertently done this publicly, thus misusing the custodial authority of the elite institutions of state. In 1814 Laplace emphasised this point in his *Philosophical essay on probabilities*, the same text in which he invented his 'vast intelligence'. There he wrote that:

> the mind has its illusions, like the sense of sight; and in the same way that touch corrects the latter, so reflection and calculation equally correct the former . . . the coincidence of some remarkable events with the predictions of astrologers, divinators and augurers, with dreams, with numbers and days which are reputed to be happy or unhappy, etc., has given birth to a host of prejudices which are still widespread.

Because 'the illusions of error are dangerous, and only the truth is generally useful', it was argued that there was thus a very close link between cosmological significance, the strategy of probability, celestial mechanics, and public welfare.

The period of Laplacian mathematical science witnessed the establishment of a defined, if fragile, community of scientists. Many of the institutions we now associate with modern science seem to have emerged during the early nineteenth century. One of them was the appearance of a public for scientific messages. The scientific community became vested with growing power: Michel Foucault observes that 'from the nineteenth century on, every scholar becomes a professor or director of a laboratory'.

Accompanying these institutional forms, a scientific utopianism emerged in which Laplace's dream was reinforced by the modish term 'social physics'. The phrase is due to Henri Saint-Simon, a student at the Polytechnique whose social theory directly used Laplacian physics to model the relation between intelligence, physics and social power. Picturing a dramatic utopia in 1802, Saint-Simon paraphrased the Laplacian dream: 'just suppose that you had acquired the knowledge of how matter is distributed at some particular time,

*Figure 7* The passage of a comet near the Big Dipper: from Grandville's *Un autre monde*, Paris 1844

and that you had made a plan of the universe . . .'. He used this model to build an ideal society, a Newtonian one, in which a 'Newtonian guard' dominated culture and in which 'the mausoleum of Newton' would operate as supreme council and scientific institute. 'I have placed Newton at my side to control enlightenment and command the inhabitants of the planets.'

This formation of a scientific utopianism, a social physics and a centralist politics was a characteristic of Laplace's culture. That 'vast intelligence' of which Laplace spoke was made the ideal typical form of scientific knowledge. This was the kind of knowledge which John Herschel used in his speeches on comets and their uselessness, and this kind of science still provides the ideal for futurology.

### 'A MIND WHICH KNEW ALL THE ATOMS IN THE UNIVERSE'

These may seem bold claims to make for a question as limited as the return of comets. But the purpose of my story has been to emphasise the way in which the security of prediction is linked with the status of the predictor. I have argued that the security of a predictive science seems to depend on a special connection betweeen the insulation of the scientific community and its grip on public interest. Tycho forged a new account of the place of the courtly astronomer, secure within his observatory but of indispensable significance for the state. Newton and his allies helped transform, rather than subvert, public investment in the meaningfulness of the heavens, turning its messages to politico-religious purposes.

Lalande, by contrast, failed to engineer this connection. He breached the boundary between the Academy and the plebs with disastrous consequences. Laplace and his colleagues sought to manage a more disciplined formation in which intellectual expertise was guaranteed a managerial role in society. Laplacian cosmology banished cometary collision to the realm of probabilistic calculation, and then devoted itself to the education of the public in the meaning of such terms as chance. In so doing, Laplace proclaimed not the end of physics but the appearance of the mathematical physicist as a new ideal mind. Ian Hacking has written of this nineteenth-century process as the 'taming of chance'. I have argued here that this process of taming was also productive for the contemporary notion of the expert, and that in fact the status of scientific prediction in general, and of cometary prediction in particular, goes hand-in-hand with the emergence of this new role of 'scientist'.

Look at the imagery deployed in one of the most celebrated

nineteenth-century formulations of determinism, that of the great German physiologist Emil Du Bois-Reymond, speaking to a rally of German natural scientists and medics in August 1872:

> A mind which knew for a given very small period of time the position, direction and velocity of all atoms in the universe, would be able ... by an appropriate treatment of its world-formula, to tell us who was the Man in the Iron Mask ... As the astronomer predicts the day on which, after many years, a comet again appears in the vault of heaven from the depths of space, so this 'mind' would read in its equations the day when the Greek cross will glitter from the mosque of Saint Sophia, or when England will burn her last lump of coal.

Hacking notes that this statement has often, mistakenly, been seen as the inauguration of a modern notion of determinism, that all events have determinate causes and are therefore predictable in principle. In contrast, I use it here to demonstrate that there is a very important connection between the way experts talk when they claim to be able to foresee the future and the kind of mentality which these experts claim they possess. Our culture oscillates between wild optimism about the promise of a technological utopia and dark distrust of technological expertise. In this context it is fascinating to notice how commonly talk of the end of all things, including such apposite disasters as religious war in the East and British economic collapse, is still sustained by the image of cometary prediction.

## FURTHER READING

Burke, Peter, *Popular culture in early modern Europe*, London: Temple Smith, 1978.

Capp, Bernard, *Astrology and the popular press*, London: Faber, 1979.

Curry, Patrick (ed.), *Astrology, science and society: historical studies*, Woodbridge: Boydell, 1987.

Christianson, J. R., 'Tycho Brahe's treatise on the comet of 1577', *Isis* 70 (1979), 110–40.

Gigerenzer, Gerd, Swijtink, Zeno, Porter, Theodore, Daston, Lorraine, Beatty, John, and Kruger, Lorenz, *The empire of change: How probability changed science and everyday life*, Cambridge: Cambridge University Press, 1989.

Hellman, Clarisse Doris, *The comet of 1577: its place in the history of astronomy*, New York: Columbia University Press, 1944.

Herrmann, Dieter B., *The history of astronomy from Herschel to Hertzsprung*, Cambridge: Cambridge University Press, 1984.

Jaki, Stanley, L., *Planets and planetarians*, Edinburgh: Scottish Academic Press, 1978.

Merleau-Ponty, Jacques, *La science de l'univers à l'âge du positivisme*, Paris: Vrin, 1983.

Taton, René, and Wilson, Curtis (eds.), *Planetary astronomy from the Renaissance to the rise of astrophysics, Part A: Tycho to Newton* (General History of Astronomy, Volume 2A), Cambridge: Cambridge University Press, 1989.

Thrower, Norman (ed.), *Standing on the shoulders of giants: a longer view of Newton and Halley*, Los Angeles: University of California Press, 1990.

Webster, Charles, *From Paracelsus to Newton: magic and the making of modern science*, Cambridge: Cambridge University Press, 1982.

4
———

# Predicting the economy

*FRANK HAHN*

How effectively in the 1990s can we predict the future course of a country's economy? In analysing this key issue in the chapter that follows, my aims are essentially threefold. First, I will consider the various meanings of prediction and note some of the special problems that economic prediction presents. Second, I will argue that there is a distinction between understanding and predicting, economics being significantly better at the former than the latter. Finally, I will consider the claim that the natural and social sciences are so essentially different that we cannot expect predictions from social science.

## PREDICTING

Every non-tautological proposition concerning the world restricts what the world is like. In that sense all such theories predict, and economic theory is no exception. The temporal aspect here is not essential. When, for instance, a physical theory predicts the existence of a new particle, there is no reference to the future. When an economist predicts that economies with price controls are highly likely to have black markets, no reference to time is needed.

A substantial part of economic theory is of this kind and we refer to the analysis as one of comparative statics or dynamics. Thus we may consider and compare the equilibria of two economies which are identical in all but one respect: one, for example, may have a tax on tobacco while the other has no such tax. If we then deduce something about the relative prices of tobacco and, say, beer in the two economies we are not predicting the future. What we are predicting is what would be found if two actual economies satisfying the assumptions were to be compared.

This method will be familiar to experimental biologists. The advantage biologists have, however, is that they can be a great deal more certain than economists that, when they come to compare a control group with an experimental one, the conditions of the comparison are fulfilled.

The possibility of experiment is the possibility of putting the restrictions a theory imposes on the world to the test by ensuring that the conditions of the theory hold. Though the scope for experiment in economics is rather limited, some experiments have been undertaken. Certain propositions of game theory have been put to experimental test by getting students to play a simple game for real money under the conditions (such as lack of direct communication) which the theory demands. The results have been useful and interesting, but plainly we may doubt whether people act in the same way in non-experimental situations. Experiments of this sort provide insights but no certainty.

More generally, economists must resort to statistical inference. Many years ago, to cite but one example, Milton Friedman constructed a rather beautiful theory to the effect that out of given receipts agents whose receipts were more variable would save a larger fraction than agents whose receipts were less variable. He then proceeded to test this theory by cross-sectional analysis: that is, by comparing the saving behaviour of various types of people at the same date. This enabled him to rule out the influences on savings of other factors – say changing interest-rates – which would have to be considered in a time-series test. He found indeed that his theory predicted correctly: that the saving ratio of farmers out of given receipts would generally be higher than that of doctors, and that of black workers generally higher than that of white workers. But these tests could not be as conclusive as experiments in physics. It was open to someone to argue, for instance, that the variability of receipts was correlated with some other factor that in reality accounted for the observed difference in savings behaviour. And, of course, there are always endless doubts concerning the reliability of data – do people report their savings correctly, for instance? Economists have to live with such doubts, anticipate them, and try to resolve them as far as the nature of the subject allows.

**Prediction without the time dimension** It is important to recognise also that not all predictions involve the future. A second economic example will demonstrate this in more detail.

Most of economics is founded on axioms defining what is meant by rational choice. To help formulate our ideas, suppose that the objects of choice are bundles of goods. The first axiom is that an agent 'knows what he or she wants', a statement formalised by saying that an agent has a complete pre-order of the bundles of goods. If we therefore consider two collections of goods, $x$ and $y$, the agent either prefers $x$ to $y$, or he prefers $y$ to $x$, or he is without any preference. A preference ordering is like any other ordering, say, of people by height: if Jones is taller than Smith and Smith is taller than Robinson then Jones, logically, must be taller than Robinson. Similarly, if of three bundles $x$, $y$ and $z$, $x$ is preferred or thought equal to $y$ and $y$ is preferred or thought equal to $z$, then $x$ is preferred or thought equal to $z$: the ordering is transitive. The assumption that we have such a well-defined preference ordering over possible consumption bundles is, of course, quite strong.

The second axiom states that the agent knows how to get what he or she wants. This is formalised as follows: given the agent's choice of goods and thus expenditure, there must be no other bundle costing as much or less which he or she prefers to the bundle chosen.

This theory leads to a number of predictions of the kind which I am now considering. I take a very simple example: let a Cambridge college have two possible ways of charging students for meals. The first

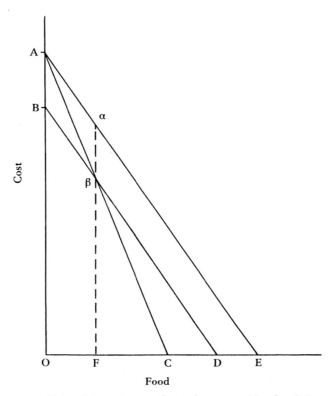

*Figure 1* A student prefers to be on BD with a fixed charge of AB
and marginal cost for food to being on the full cost line AC

(to be called 'full cost') charges a price such that, at the number of
meals sold, both the overhead and the variable (marginal) costs of
meals are covered. The second (to be called 'two-part pricing')
imposes an establishment charge equal to overhead costs and thereafter charges marginal costs. Once again no profit is made. If all
students have similar tastes and incomes and all eat some meals, then
the predictions of the theory are (a) that if the two schemes are put to
the vote no student will prefer the full-cost scheme; and (b) that under
the two-part pricing scheme students will not take fewer meals than
they would have done under full-cost pricing.

In Figure 1, at full cost any choice in the triangle OAC is available to
the student and we suppose that $\beta$ (i.e. OF food) is chosen. Had the

college only charged marginal cost, any choice in the triangle OAE could have been chosen. It follows that, when at full cost, OF food costs $\alpha\beta$=AB more than it would have done if only marginal cost had been charged.

Thus if the college charged an establishment fee equal to AB it would be sure that its overheads were covered. Charging marginal cost for any food actually bought means that the student can choose anything in OBD. But he will not choose any combination lying in OAC because he could have chosen that before and did not, and so preferred $\beta$ to any such point. If he changes his choice it must lie in C$\beta$D. Since $\beta$ is still available, any such point cannot be worse in preference than in $\beta$. Hence the student will be willing to vote for two-part pricing over full-cost pricing and will consume at least as much food under the former scheme as he would under the latter. These are two predictions of the theory which do not involve time in any essential way.

You will have noticed that I assumed that all students were alike in taste and income. If this is not so, the proposition needs to be reformulated. If a given overhead charge is added to the price, students with different consumption of food are making different contributions to overheads. The prediction now is that each student would at most be indifferent (and would generally prefer) to make this contribution in the form of an establishment charge rather than doing so by means of full-cost pricing.

**Predicting the future** The prediction of future events is, of course, of as much interest to economists as prediction without the time dimension. It will be obvious that all such predictions must be conditional. Less immediately apparent is the fact that in a subject like economics conditionals require special care.

Suppose we ask: What will the British inflation rate be next year? We all know that the answer will depend partly on the course of interest rates. If interest rates are under the Chancellor of the Exchequer's control, then in giving his own forecast we may consider interest rates as a conditional of our forecast. But if current inflation forces down exchange rates or puts great pressure on the balance of payments, interest rates may largely cease to be under the Chancellor's control. In that case, to predict inflation we must also predict interest rates; this in turn may require a prediction of exchange rates, and so on.

Strictly, then, the conditionals of a prediction should be exogenous to the theory – that is, the theory does not explain the conditionals. But that is being very strict – indeed, one might argue, unmanageably strict. Economists in their theorising are aware of the general inter-dependence of economic variables: for instance, the price of pianos may affect the demand for television sets and hence their price; this in turn can affect the demand for cinema seats, and so on. We know how to formalise such interdependence.

Put simply, our theory leads us to the view that in general the demand and supply of any one good depends not only on its own price but also on the prices of everything else. Amongst 'everything else' we include the price of leisure (i.e. wages) and the prices of goods in the future.

Let us now make two assumptions: agents have correct expec-tations of future prices and at every date prices are such as to ensure that the amount of every good agents demand is equal to the amount of the good agents want to supply. Strictly we also need to assume that at every date there is only one set of prices at which demand equals supply for every good. With these assumptions we have con-structed a dynamic system as follows.

At any one date the prices which clear all markets are uniquely given. They depend, of course, on the prices which are expected to rule in the future. But we have assumed that these expectations are correct. So if we know today's prices we also know tomorrow's prices which, because they were expected today, determined today's prices. When tomorrow comes we know what prices will be which clear markets because they are simply the prices correctly expected today. That in turn implies that we can deduce the prices, correctly expected tomorrow, to clear the markets the day after tomorrow, and so on.

Clearly, we must know all the functional forms involved here, and in general we do not know them at all accurately. They must be econometrically estimated and there are reasons for arguing that this estimation must be simultaneous and not one at a time. There will certainly be errors of estimation and in non-linear systems even very small errors may lead to widely different forecasts of the course of prices given the path of the exogenous variables. The same is true if we only know prices inaccurately. Predicting from such a complex system is therefore not easy.

As a result, when we set out to make economic predictions over time, we have to depart to some extent from what theory demands. As a first step we want to reduce the dimension of the model by aggregation: that is, instead of one equation for each good separately, we may, for instance, take one for an aggregate called consumption goods, another for investment goods, and so on. Second, our method of estimating excess demand functions mostly means that we have linear estimates. This may be less serious for what might be called normal values of the exogenous variables – assuming, for example, no earthquakes, no wars, and no revolutions – than it is for others. It is also more reasonable if our conditional forecast does not extend into the far distant future.

I am not suggesting that the theory which I have used is a good one – we can do better than that. But it suffices to illustrate the two important points I wish to make. First, in our forecasts we must at each stage state what is assumed concerning the path of the exogenous variables, and we must be sure, of course, that they are exogenous. In our case we assume that this path is independent of prices. Second, we must recognise that it is simply beyond our capacity to proceed without some more or less *ad hoc* doctoring of our theories to make matters manageable.

There is a further moral also: our predictions cannot be exact and will be subject to error. This follows from inevitable errors of measurement and errors in the estimation of functional forms. In addition one must bear in mind that the path of the exogenous variable which one takes as given may be a stochastic path.

In spite of this we may hope to predict the expected values of our variables where 'expected' is to be taken in the statistical sense. This kind of prediction will be familiar from physics. However, here, in the nature of the case, such averages must concern the characteristics of a long future path. Keynes observed that 'in the long run we are all dead', by which he meant that we should not even bother to pronounce on the long run. If, then, we concentrate on the short run we cannot appeal to averages – as we sometimes can if the stochastic process of the exogenous variable is stationary – and accordingly will often find large deviations. On the other hand, we will have more certainty about the exogenous variables since in many cases they will move slowly relative to the main variables of interest. In short-run predictions there are reasons to believe that we escape some of the notorious surprises which non-linear dynamics may have in store for us and the very large errors of prediction which may result from errors of measurement.

All of this may lead to a rather pessimistic view of what economists can hope to achieve. Later in the chapter I will discuss some examples of what we can nonetheless hope to predict; but first I want to draw a more general lesson.

The desirability of prediction rests on two rather different considerations. One is that prediction is designed as a test for theory, the other that it serves as an aid to action and decision. I will argue shortly that even false theories may satisfy this second desideratum. As far as economics is concerned I hope I have also said enough to show that there are very large difficulties with prediction as a test for theory. Ultimately these stem from our inability to experiment scientifically, that is, adequately to control for the exogenous variable and for the intrinsic interrelatedness of economic phenomena. It has therefore proved hard to obtain decisive rejections of economic theories, a fact which means in turn that we must rely in the first instance on internal theoretical coherence in any evaluation.

## UNDERSTANDING

Before returning to further discussion of economists' power to predict I want to take two small detours. The first of these concerns the function of theories to further understanding.

It seems to me that the ability of theories to predict is neither a necessary nor a sufficient condition for understanding that part of the world which a theory covers. Examples are easy to come by. Most geologists – perhaps all – are agreed that the theory of plates provides an adequate explanation for earthquakes. Yet though we understand earthquakes, they cannot at the moment be predicted. Similarly, there are certain non-linear dynamic systems where the equations reflect our understanding but prediction is nonetheless impossible. Evolution, of course, furnishes another example. It seems therefore that predictability is not necessary for understanding.

The lack of sufficiency is provided by any of a large number of 'nonsense-correlations'. For instance, two of my colleagues found that the incidence of rickets in Scotland was better correlated with the price level than are money aggregates. So if we can predict rickets we can predict the price level, but why this should be so would be pretty mysterious. We would be predicting without understanding.

In these examples I have made no distinction between predictions which do, and those which do not, involve time in an essential way. Thus geophysicists may be unable to predict earthquakes and yet find it possible to calculate the pressures which, if they occurred, would precipitate a quake. This possibility I did not wish to exclude. My point is that even if such calculations prove impossible, scientists are still entitled to claim that they understand earthquakes. Understanding always involves prediction in the sense of my opening remarks: it places restrictions on what the world can be like.

The notion of 'understanding' is by no means straightforward and Wittgenstein, for instance, had a number of somewhat poetic things to say on the matter. Not being a philosopher I will, I hope, not be expected to propose a watertight definition. I will explain what I mean but will not be surprised if philosophers find my meaning unsatisfactory.

I hold that a theory which provides understanding necessarily has two characteristics: (a) its statement involves no logical error; and (b) it does not contradict – indeed it fits in with – other things we think we know. I believe that (a) is not controversial. But (b) is both loosely formulated and more difficult. For instance, Einstein plainly thought that its requirements were not satisfied by quantum theory and, reading accounts of the debate, he was not alone in this. For an outsider it is certainly hard to reconcile the theory with the axioms of logic which we think we know and need. Apparently physicists themselves have difficulty in reconciling the theory with Einstein's theory of gravity, so perhaps one would be justified in claiming quantum theory to be only partly understood. But one sees how hard it is to apply my second criterion.

Now consider an example from economics. Some economists think of an economy as a single, infinitely lived and perfectly informed, representative individual, living in a stochastic world. The stochastic process is stationary and the representative agent maximises an infinite sum (integral) of his expected utilities, where utility at each date and realisation of the process depends on the consumption of a single good and the amount of work done. The maximisation proceeds under the stochastically given opportunities of production. The result is an optimal stochastic path of output and employment. Empirical implementation is provided by an estimate of the utility function and of the constraints. Many things could be said concerning the way this is accomplished, but this is not the place. For a time it was claimed that the results gave a very good fit and so could be used predictively. Whatever the merits of this last claim, the essential question is whether the theory provides understanding.

Certainly it does not fit with other things which we know. Actual people have finite lives. We know that people often make mistakes in their economic forecasts. We know that markets are not generally perfectly competitive. We have strong theoretical grounds for believing that markets do not clear at every instant, and so on. But the negation of all these things is essential to the theory which I have described, so one seems justified in saying that one does not understand why the theory should predict well on occasions when it does so.

There is here an instance of a profound difference between the way American and European economists view their theories. It is best exemplified by a very influential methodological piece by Milton

Friedman which argued that predictive success is the sole test of a theory. For instance, in the example I have just given we should understand the theory as saying that the economy behaves 'as if' it were populated by a single infinitely lived agent with perfect foresight, etc. If prediction is good therefore, no more needs to be done. For others, including myself, this is unsatisfactory. We want to understand why the 'as if' assumptions are successful. But we are also aware that in a non-experimental subject where there is no tight control over conditionals there must be small confidence in a unique 'as if' theory. The methodology allows far too much latitude. Lastly there are logical difficulties. The assumptions are themselves theories. For instance, the theory that prices at all times 'clear' markets is a poor predictor. Would not Friedman's methodology lead one to reject this and so also the larger theory of which it is a constituent part?

I have no difficulty in rejecting the pseudo-scientific posture of the 'as if' methodology, although it will have been noted that I am myself vulnerable because my notion of understanding is so loose. I cannot see my way to a remedy. However, a little tidying-up is possible before I leave the matter.

There are theories which summarise some empirical regularity: taking aspirin, for instance, will relieve headaches. Theories like these are purely predictive and they have only an elementary logical structure. As criteria for understanding they have no bite. Yet the sentence 'We do not understand why aspirin relieves headaches' is immediately comprehensible. This suggests that we should add a third criterion in order to arrive at conditions which are both sufficient and necessary for a claim to understanding. This must be that they leave unanswered no 'why' or 'how' questions beyond some basic epistemological axioms. In this way Newton's theory which involved action at a distance was claimed by some as providing prediction without understanding, and did indeed act as one impulse to further research and to reformulation.

While I am pretty confident in the distinction between prediction and understanding which I am making, I have failed to define the difference precisely. To do so is a task for philosophers. I am also uneasy with the inevitable subjective element in understanding. Prediction is a public act; understanding seems to be more private. What we claim to understand depends on what we think we know and what we think we have already understood – and both types of knowledge are subject to change.

All of this is true. Yet I wish to claim that economists do understand a great deal about decentralised economies without having theories of great predictive success. Adam Smith provided an understanding of how an economy with self-interested agents could be orderly, an intellectual achievement of a high order. But this 'understanding theory' does not allow us to predict that any actual economies will be orderly in practice.

## OBJECTIONS TO THE POSSIBILITY OF SOCIAL SCIENCE PREDICTIONS

I now want to consider the argument associated with Max Weber that social scientists cannot predict as a matter of logic. Recall that for this reason Weber distinguished between *Natur-Wissenschaften* and *Geistes-Wissenschaften*, the natural and the social sciences. I believe that his reasons and some others which have been adduced are incorrect. First I consider Weber's reason which I shall call 'contamination of the predicted by predictions', second the objection from 'free will'.

**Contamination of the predicted by prediction**  Suppose there is to be an election between two parties – say Labour and Conservative†. A social scientist attempts to predict the fraction $x$ which will vote Labour. This prediction is known and it will affect the proportion $y$ who will actually vote Labour. There is a bandwagon effect, and from this it is deduced that no correct prediction is possible.

Many will see at once why this conclusion does not follow. In the

---

† My attention has been drawn to a classic paper which inexplicably was unknown to me, and which proceeds as I do here: E. Greenberg and F. Modigliani: 'The predictability of social events', *Journal of Political Economy* 62 (1954), 465–78.

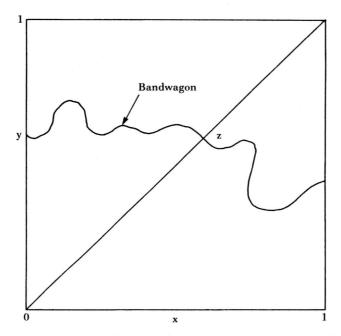

*Figure 2* The bandwagon curve must cross the diagonal and/or lie on it at the corners

diagram above $x$ represents the proportion predicted to be going to vote Labour while $y$ is the proportion which actually does so. The bandwagon curve shows how actual voting is affected by the prediction. It can be any old shape but we demand that it be continuous – we can draw the curve without ever taking our pencil off the paper. At $Z$ where the curve crosses the diagonal we have $x=y$, that is, a correct prediction.

However, the curve must cross the diagonal at some point. If the curve started at $y=0$ then $x=0$ and '0' is the correct prediction. If the curve meets the diagonal at $y=1$ then once again $x=1$ and the prediction is correct. If neither of these possibilities occurs then the curve must cross the diagonal at least once. If it stayed above the diagonal for all $x>0$, it would have to equal $x$ at $x=1$ contrary to assumption; if it stayed below the diagonal for all $x<1$, then it would have to equal $x$

at $x=0$. (This is a consequence of continuity.) Hence a correct prediction is always possible.

We have used an elementary fixed point theorem in this argument, but there are many such theorems for all sorts of complicated cases and economists have made much use of them. There is in any case no *logical* entailment of an impossibility to predict which is deducible from 'contamination'. This does not, of course, mean that we can predict correctly in practice. We would need to know the bandwagon function.

**Free will** Human beings are not physical objects. They are autonomous agents with their own projects, and their actions are by their very nature contingent on their will. Reductionists may deny this but I am happy to accept it. What I deny is the proposition that the ability to predict an agent's action is equivalent to denying that agent 'free will': that is, denying the possibility that he could if he so wills choose otherwise.

The argument here seems to me pretty obvious. An agent may choose otherwise than predicted, but he does not will it. Suppose that there are two ways of milking a cow, one taking twice as long as the other. If the farmer is free to choose either, I predict that he will choose the more efficient because he values time. This prediction turns out to be correct because the farmer does indeed value his time. Unlike the case of an object, there is no necessity here; there is no law. It just is the case that we can predict the farmer's choice here, today, in England. Exactly the same argument can be used for my example of the pricing of student meals.

Economists would in practice be a little more circumspect. They would take account of the possibility of stupid or ignorant farmers and of the possibility that a farmer might, for the sake of variety, sometimes choose a less efficient method. Our prediction would be in the form of a probability that any given farmer is an efficient milker. This is particularly true of more complex choices than in the example.

Suppose next that there exists a profit which can be made by arbitrage between two currencies. We would predict that arbitrage would indeed occur with a high degree of certainty. We are not claiming that genetic or other mechanical forces are responsible. It is perfectly possible for agents to choose not to take advantage of the profit opportunity; but they will do so not because they have to but because they choose to do so.

All of this seems clear enough and I will not labour it further. However, while the argument from free will I have been considering is just wrong, there remains a different implication which we ought to take seriously. This is that we must expect very few 'laws of economics'. By this I mean no more than my earlier observation that in economic analysis we will always need to take account of exogenous variables such as history and social norms.

Such 'laws' as economics can provide are rather trivial, although it is surprising that they are not widely understood. Thus, for instance, it is a law – it is really a physical law – that where leisure is included amongst goods, we cannot in general at any one instant have more of one good without having less of another. But when one passes beyond such trivia, exogenous and thus unexplained variables cannot be ignored. We hear of 'the law of demand and supply'. But there is no such law which holds in all societies at all times. For instance, early in the last century when there was an excess supply of labour there was indeed a tendency for wages to fall; this is a far less certain response in the age of trade unions.

So perhaps Weber after all had a point with his distinction between *Natur* and *Geist*. One must suppose that there really are 'laws' of physics. To suppose that there are also laws in the social sciences requires us to assume that we shall find a closed theory: that is, one with no exogenous variables. This has been tried – Marx comes to mind – but the attempt cannot be claimed to have been a conspicuous success. When it comes to grand closed theories one cannot precisely argue that such theories must be in contradiction to human autonomy and free will, but one suspects, none the less, that they will be bogus. Complexity sometimes implies impossibility. Sir Karl Popper maintained that such theories were intrinsically impossible; I rather think he was right.

## LAST REMARKS

It will be clear from all that I have argued that if there is a natural science which has similarities to economics it is biology, not physics. It is easy to translate many ecological theories into economics. Similarly, many biologists have used the economics of cost–benefit analysis and game theory to gain insights into evolutionary processes and ethology. Biologists and economists also have in common an ability to provide a good deal more understanding than prediction. Some of the most exciting work in economics at the moment is rather close to evolutionary theory. Regarding the market as a selection mechanism, for instance, is proving fruitful and capable of precise formulation. This has the advantage that we have a further reason for not agreeing to objections from free will quarters; the survivors are survivors because they have chosen to act in a certain way.

The evolutionary approach suggests also that there is no historical necessity for the economy to be as we find it to be, just as it would have been possible to find a different array of species than we now have. No one expects biologists to predict *Homo sapiens*.

All of this is at a considerable level of abstraction and you may therefore be thinking that economics has nothing concrete to offer when it comes to prediction. I have already given several examples to persuade you of a contrary view, and I will conclude with a last – rather topical – case.

We in Britain want to know some of the main consequences of joining the European Exchange Rate Mechanism, the ERM. Theory provides a safe short-run prediction: it will increase unemployment and reduce output. The argument is quite simple: profit-seeking agents will see to it, given our inflation rate, that the pound sterling will have a tendency to fall through the lower part of the exchange rate band. Interest rates must therefore be high enough to counteract these tendencies. Of course, the government may not be willing to maintain sterling's position in the band: so we have a prediction, but one conditional on the British government being seriously committed to the existing value of sterling.

Part of the argument here rests on the recent history of inflation and of what we know of the labour market. For money wages not to compensate for past erosion of real wages is unlikely as long as unemployment has not risen. Once again this is not a certainty, but both theory and empirical knowledge point to it being the case. All of this then leads to an inflation rate higher than on the continent and the predicted unemployment follows. But by how much will unemployment rise? Here we can provide estimates, but for all the reasons I have discussed, they will be estimates that embody considerable errors. On the other hand, we can say that the higher the rate at which we entered the ERM, the higher unemployment is likely to be.

When we come to the longer run, our predictions become less sure. There will certainly be a limit to the inflation and the rise in unemployment if the government continues to be seriously committed; again we can give estimates. However, we will be less sure of the time which will be taken. Even harder still is deciding whether we shall then be saddled for an extended period with high unemployment and low inflation. In this case a great many new considerations come into play. First, we need to make assumptions about government policy. Suppose it is felt safe to lower interest rates. We must then discuss what will happen to investment and exports and imports. It is easy to show that it is possible for inflationary troubles to recommence – lower unemployment accompanied by too rapidly rising money wages, etc. One can also show that it is possible that this may be avoided. To predict which will happen is, I believe, only possible with the eye of faith.

There is, however, a silver lining. Economists are often able to argue that certain policies make one or the other outcome more likely. That is a sort of prediction and it is useful. What occurs when we pass to the more distant future is that economics provides a fairly powerful tool for evaluating alternative policies; by that I mean that it allows us to discuss what must be the case to change the likelihood of possible outcomes.

All of this is rather far removed from the certainties of politicians. But when one considers the complexity of economic life I don't think that it is to be despised. Economists provide a grammatical way of peering into the future even if they do not always see it clearly. This is better than consulting entrails.

## SUMMING UP

The most important proposition I have sought to put to you is that economics is pretty good at the business of understanding and at providing a grammar of argument. That makes it a useful discipline even if its predictive powers are rather limited. There is not, in my view, anything unique and fundamental in the nature of the subject matter that inevitably makes economics inter-temporarily non-predictive.

I have criticised two of the most frequently heard arguments, from free will and contamination by prediction. The main obstacles in the way come, first, from the complexity which inevitably introduces exogenous or unexplained variables and straightforward computation problems into all economic theories. Second is the necessity of relying on statistical inference rather than experiments to discriminate between theories. Third is the problem that our data are never precise enough to escape the large errors which we know can be generated by non-linear dynamics. Yet even faced with these difficulties, I have argued that we need not despair of getting short-period forecasts reasonably correct.

I can explain by an example what I mean by there not being something peculiar in the nature of the social sciences which is bound to make them unpredictive. Aerodynamics is an impressive body of precise theory. It is certainly science. Yet no aeroplane designer would be prepared to predict the performance of a new design without wind tunnel and other tests. The complexity the designer has to contend with is, I suggest, several orders of magnitude smaller than that found by the economist dealing with the problems of his own field; but the economist has no wind tunnels. There is nothing deep and fundamental here – just a contingent limitation. The surprise is not that we predict so badly but rather that, in the relatively short term, we predict so well.

There have been discussions of whether economics is a science or not. If science means the serious, disciplined study of observed phenomena, it clearly is. If, however, science means uncovering 'laws' and providing accurate inter-temporal predictions, I should say that it is not. My impression is that most people interpret science in this latter way: good reason therefore for economists not to lay claim to the mantle of science. When they do claim it, they sooner or later babble about 'the laws of economics' and begin to say much more than it is prudent to say. The subject gets a bad name. Yet economics has in fact cast much light and it can be a powerful weapon against nonsense. Certainly if it did not exist, it would urgently need inventing.

## FURTHER READING

Geanakoplos, John, 'Arrow–Debreu model of general equilibrium', in *The New Palgrave dictionary of economics*, ed. John Eatwell, Murray Milgate and Peter Newman, London: Macmillan, 1987, Volume 1, pp. 116–24.

Nikaido, Hukukane, 'Fixed point theorems', in *The New Palgrave dictionary of economics*, ed. by John Eatwell, Murray Milgate and Peter Newman, London: Macmillan, 1987, Volume 2, pp. 386–7.

Varian, Hal R., *Micro-economic analysis*, New York and London: Norton, 2nd edition, 1984.

Varian, Hal R., *Intermediate micro-economics*, New York and London: Norton, 2nd edition, 1990.

# 5

## The medical frontier

*IAN KENNEDY*

This book's aim is to consider the underlying theories of knowledge on which predictions are based and changing conceptions of the purpose and role of predictions, and I am invited to consider in this context an issue that is today of universal concern: the frontier of medicine and the new legal and moral dilemmas that emerge around it every day.

True to my training as a lawyer, I will endeavour to keep faith with this brief. As a consequence, I will not be inviting you for a rapid gallop through all the wonders and horrors of modern medicine with some instant prescriptions, such as 'there ought to be a law against this or that'. Instead, I will offer what at best can be interim thoughts on matters self-evidently of great complexity.

Consider, as a starting point, four dilemmas that may not be so far in the future:

1. In the field of human genetics, (a) it may soon be possible to test for susceptibility to a wide range of disorders both monogenic and polygenic. An employer informs all workers and those applying for employment that he requires them to undergo testing. He will not employ those who are not a good investment, or those who, if he pays employees' health insurance, will generate large medical bills in the future. Some find themselves unable to obtain employment. Alternatively, (b) a register is kept of human genetic information.

A record of the genetic make-up of a particular individual is held on this register. Agencies of the state, or employers, seek, in what they argue is the public interest, access to this information, although it was given on terms that it be held privately. In the same way, (c) it is proposed to carry out not merely somatic cell gene therapy resulting in non-heritable changes to body tissue, but germline therapy whereby heritable genetic modifications are made.

2. No one knows how many patients there are in the United States in a persistent vegetative state. It is estimated, however, that the number runs into many tens of thousands. The cost of maintaining each of these patients is many thousands of dollars. Given that their prognosis is hopeless, it is proposed to regard them as dead with a view to removing vital organs to supplement the supply of such organs for transplantation.

3. Resources available for health-care have become so scarce as a consequence of their prohibitive costs and the virtually unlimited demand upon them that hospitals ask for police protection and armed barricades to prevent access to any but the wealthiest, or close geriatric wards to divert money to kidney dialysis and neonatal care units.

4. As a consequence of global warming, revolution, or war, more than 50 million people, most of whom are in need of basic medical care, move into Western Europe.

These are a few hard dilemmas. They are, of course, dilemmas of the developed world, which is the world that I shall concentrate on; the frontiers are elsewhere in the Third World.

It is easy enough to identify these dilemmas: to predict at the level of hypothesis. But, as I have said, it is not my purpose to spot the dilemma, say 'Gee whiz' and move on – even though this is a practice much loved by some commentators. My task instead is to consider how we will respond. My submission is not startling: our responses will be determined by the way in which we perceive the dilemmas in question.

## THE TRADITIONAL APPROACH

To discover what responses we will make, it may be helpful first to notice how hitherto we have tended to perceive medical dilemmas. Put crudely, our ways of perceiving them and hence our responses, have been the product of the triumph of medicine in setting the mode of discourse. Medicine has in the past been assumed to be a value-free enterprise engaged in the conquest of illness – a view with which non-medical people as much as doctors have been in complicity, hoping it to be true. The technological imperative, that if it can be done it should be done, has prevailed.

From this and other such perceptions has grown the view that the story of medicine is of an ineluctable process of development towards the conquest of disease. Since the conquest of illness and disease is a 'good thing', medicine and medical developments are 'good things' and those practising medicine are by extension 'good things' also. Responses to our dilemmas born of this tradition will take the form that medicine should continue to develop. Predictions about the future will be shaped by how we have perceived medicine in the past. On this view it is not the responsibility of medicine if nasty problems arise from time to time. In any event, if they do, they can safely be left to doctors, imbued with a true understanding of the enterprise of medicine, to deal with. Only if they are particularly nasty, will someone else have to do something.

The evidence sustaining this mode of discourse was, of course, always somewhat fragile. As regards the conquest of illness, the great killers of our age – poverty, tobacco, alcohol, motor cars and pollution – remain largely untouched by medicine. Equally, the bringers of ill-health – poor housing, ignorance, social isolation and unemployment, all of which stunt the growth of individuals and society – also remain. But, as you know, because medicine is power-less here, it simply re-defines its concern as disease rather than health and hurries on.

Within medicine, the triumphs of technology have produced mixed blessings. The capacity to extend care to new-born babies has brought joy but also the anguish of deciding which babies should benefit and which should be left to the whim of nature. The capacity to effect marvellous changes in the sick by transplanting tissue led to the implanting of a baboon's heart in a baby girl, mechanical hearts into very sick men, and calls to regard anencephalic babies as dead so as to harvest their organs.

One particular claim of this mode of discourse merits notice as in many ways it was the jewel in the crown of medicine. Infectious diseases were a thing of the past we were told, at least in the developed world; they had been all but conquered. Then there was rather a rude shock. A new plague appeared – AIDS – just to show that there is life in the old bugs yet and to remind us of the penalty for hubris.

Despite sniping attacks from the counter-culture, the perception of medicine which I have portrayed remained dominant, and an unquestioning, unreflective frame of mind was fostered. The story was taken at face value. Medicine was an inevitably good enterprise, things were good and could only get better. The future would be more of the past. If there were any problems, they could safely be left to those involved in medicine who would iron out the social and moral creases. Grandma could be given her Brompton Cocktail, the mongol baby could be smothered with a blanket.

## AN ALTERNATIVE APPROACH

Gradually, meeting resistance every inch of the way, there has of course been a change in the intellectual rules of engagement; reflection outside medicine, and gradually within, has caught on. The reasons are many and complex: an educated population, a distrust of elites, the emergence of consumerism, the extraordinary growth in the dissemination of information, an increasing awareness of the limitations and risks of medicine, and so on. The consequence has been that the responses to the sort of dilemmas I have outlined have come to be seen as calling for an entirely different frame of reference from that previously employed. Another basis of prediction has appeared. Predicting what the future of medicine will hold has come to be seen as dependent on choices, largely moral choices made beforehand: choices arrived at as a consequence of considered reflection.

Enter therefore the infant discipline of medical ethics, or bioethics as I shall call it. In the past, reflection on the dilemmas of medicine was after the event, and prediction was done when the result was known! Bioethics therefore had all the importance of the person following the Lord Mayor's Show. But now an almighty struggle has been joined. Bioethics is saying that *it* should play the leading role in setting the agenda for the future of medicine, at least as regards the big questions. It should set the terms of reference, the mode of discourse for our responses to the dilemmas at the frontiers of medicine, and determine our predictions. Predictions, the argument goes, must not be seen as the unravelling of some inevitable process of discovery and development in which medicine is value-free. Rather, they must be seen to involve the making of choices, of considering now what the future may be. What we predict, therefore, is what we prefer or consider to be good.

It follows from this that the evaluations which inform such choices are not matters of nice scientific fact but of uncertain moral enquiry. Nor are they the preserve of medicine. They are for philosophers, for theologians, for lawyers perhaps, and for the person on the Clapham omnibus. Their purpose is to arm us to make plans, to consider carefully the society and the medicine we want, to assess what may be and to ask whether it is good or bad, to be permitted or not.

There is, of course, a rearguard which still mounts attacks on this way of approaching the future. Largely it is sniping. Medical developments, it says, are value-free, and the pursuit of knowledge is an essential and eternal good. We would never have had aspirin, it asserts, if it had had to be licensed now. Innovation will be stopped in its tracks: look at the time wasted debating such issues as research on embryos, using fetal tissue for transplant or declaring a moratorium on genetic engineering. But this is, I repeat, sniping. The days of the die-hards are numbered.

The truth today is that predictions as to how we will confront medical dilemmas at the frontier depend on the mode of discourse or means of analysis employed. The appropriate and valid mode is that which recognises that what we face in the future depends on value choices made now. In short, we determine our own dilemmas.

## MEDICINE AT THE FRONTIER

Let me now turn back to the dilemmas at the frontiers of medicine. If this is taken to mean the factual prediction of what is just around the corner, you are in for a disappointment. I have no crystal ball. I have also to remind you that predicting future events is a pretty fraught exercise. I could have consulted the best brains in bioethics twelve years ago and asked them to name their 'issue of the 80s', and none of them would even have mentioned AIDS. All I can say, therefore, is that medicine and medical science are both extremely dynamic; so dynamic in fact that the frontiers are not far away, but here before us. Let me illustrate this with some problems which recently crossed my desk in the space of one week.

1. I received a letter from a woman asking my advice in a dispute with her employer. She had sought sick leave so as to be able to attend a clinic for *in vitro* fertilisation treatment. While her employer was prepared to grant time off with pay to attend ante-natal care should she become pregnant, her application that her attendance for IVF treatment be treated as sick leave was rejected. After representations, she was advised that inability to work as a result of IVF treatment qualified her for sick leave, but inability to work so as to obtain the treatment did not. At one level this is just a dispute about an employment contract. At another, it raises fundamental questions about the nature of illness and whether infertility is illness, and about the boundaries and limits of the concept of treatment, leave aside the rights of women to found a family.
2. I was telephoned by a person with a slightly odd voice whose name I did not catch. The reasons soon became clear. It was a male-to-female transsexual who, as a male, had married and fathered two children; he and his wife had divorced and he had been granted

101

unlimited access to his children. But this was before he had undergone 'sex-change' surgery. Since the surgery, his wife was denying him access, which hurt him greatly. If, however, he was to apply to court to have the access order enforced, the judge might take a dim view of his changed role and deny him any access at all. Since I had written on the legal problems of transsexuals, what did I advise? What better example could you have of legal niceties rubbing up against social realities? What better illustration could there be of the interrelatedness of the medical, the moral and the social? What more poignant appeal to human rights than the right of access to one's children?

3. I was sent a draft paper outlining plans to carry out a systematic audit of clinical performance in a particular health authority. If the audit were to be properly performed, the paper argued, access to health details, including information identifying particular patients was essential. The notion that such access would be a breach of confidence by virtue of being a disclosure of a patient's confidences to someone other than his doctor was brushed aside; the information would only be read by other doctors, and since all doctors are obliged to observe secrecy, there was no breach of confidence. This is, of course, a staggeringly nonsensical argument. If A tells Dr X something, A does not licence him to tell Dr Y unless Dr Y is involved immediately in A's health care and needs to know. In the context of audit, Dr Y could be the very person A does not want to read her health records. A may have had an abortion and not told her husband. Her husband may be the very Dr Y who is carrying out the audit!

There are, of course, better arguments in support of audit. The principal justification may be that it is for the public good, and even prospectively the individual patient's good, to carry out audit because the aim is to improve the overall quality of health care. Such a benefit may warrant as its price the occasional breach of confidence. For me, this argument is less than persuasive. Once it is conceded that the trust placed by a patient in a doctor can be routinely breached for the public good, the doctor–patient relationship is irretrievably harmed. The only solution can be that patients must be advised that their records may be used for the purposes of audit and asked for their agreement. The vast majority would, of course, consent since the benefits are obvious.

4. A doctor came to see me. Her patient was a mentally handicapped young woman. The patient's brother was severely ill and needed a bone marrow transplant. The brother's doctor was pressing the doctor to agree to the removal of bone marrow from the woman, yet clearly the woman could not consent to the intervention. She would not understand what was happening and would be bewildered by, and resent, the pain it would cause. Without the bone marrow the brother could, and probably would, die. It would be so easy to do the operation; it would be over in no time; and if the woman were competent she would undoubtedly agree. But she is not competent and others must therefore protect her interests. She appears to love her brother dearly and would probably be devastated by his death. An argument can therefore be advanced that it is in her interests to remove the bone marrow. But, is this not mere sophistry? Why not be brutally honest? Her brother's life is at stake and she is mentally handicapped. Why not take what is needed and live with the minor inconvenience caused to her?

As with the other three problems there is no obvious answer. It is clear, however, that close to the heart of the problem is our respect for those who are incompetent, whether through mental handicap, mental disorder, senility or other cause. They can be regarded as citizens like the rest of us, but requiring special protection because of their vulnerability. Or they can be seen as somewhat less than us, only partial citizens possessed of fewer rights, or of rights which can readily be waived in favour of those of others.

By these four examples – and there are many more – I hope I have persuaded you of the dynamism of medicine and the way new frontiers are constantly encountered.

## PROBLEM AREAS FOR THE FUTURE

Let me now turn again to the future. Despite what I have argued about the folly of problem spotting, we can, I think, validly argue that there are areas of medical development which warrant our attention immediately because they are foreseeably problematic if a plan of action is not worked out in advance. Such a plan will not, of course, mean that they cease to be problematic. It will merely mean that some problems may be avoided and we will have a framework for response. The unpredictable will always remain.

What are the areas that will particularly press in on us?

1. An ageing population will pose increasingly complex questions about the proper care of the elderly, the treatment of the incompetent, the role and status of Living Wills, and the care of the dying.

2. Reproductive medicine will continue to shoulder its way on to centre stage. There will be claims of a right to bear children as reproductive technology develops. Maternal–fetal conflicts will become more obvious and urgent as monitoring during pregnancy allows us to know more, and causes some to press the claims of the fetus to the point of requiring the mother to undergo treatment on its behalf – even treatment as intrusive as a Caesarean section. The very notion of the family will be open to redefinition as artificial reproductive methods, including surrogacy, develop.

3. As developments in human genetics proceed apace, we will have to decide whether gene therapy and genetic screening may bring with them a tyranny of knowledge which forces upon us choices for which none of us is yet prepared, spiritually or intellectually. One such choice might be whether to test a child to determine whether it carries a defective gene, thereby critically affecting the child's perception of itself and any future decisions it may make concerning marriage and raising a family. Another might be what public policy to adopt in the face of a decision by an employer to require potential employees to undergo screening to determine whether they would be 'good health risks' for employment. Such screening would render some people unemployable because of a susceptibility to disease which may not manifest itself in practice, or which may only develop many years in the future, by which time an effective therapy has been developed.

4. The challenge of devising strategies for distributing scarce resources will grow keener as medical technology becomes increasingly expensive and it comes to be acknowledged that medical care

resources are finite while demand is virtually infinite. Hard choices have already been made and more will be required in the future.

5. Access to, and control of, information about patients and their medical care will be increasingly problematic. I specifically mention access to information because, if information *is* power, access to it represents a paradigm of the power struggle which inevitably characterises the relationship between professional and client – especially doctor and patient – a struggle in which in the UK the patient has consistently come off second best.

6. Recourse to litigation in the aftermath of medical mishaps will trouble us increasingly unless and until alternative methods of establishing accountability and securing compensation are developed. An associated development will be the perceived need to control and regulate the conduct of medical research, especially as regards the consent of research subjects.

7. Finally, the care of the vulnerable, the mentally ill, the mentally handicapped, the AIDS sufferer, the elderly and the poor, will recapture our attention after a period of neglect in which, in Britain, a combination of lack of interest, ideology, and consequent redistribution of resources has reduced their care to a level unimagined at the time the National Health Service was set up.

These, then, are some of the principal issues which will dominate the agenda of bioethics. I emphasise again that they are issues specifically related to medicine, and to medicine in the developed world. Health will have a different agenda turning on wider arguments about what is health-damaging or health-promoting, whether it be the quality of our food, the disposal of nuclear waste, or transport policy. I also emphasise that they are the issues of the developed world. The Third World will have a different agenda, one dominated in large part by the twin catastrophes of Third World debt and AIDS.

## PREDICTIONS

What predictions can we make as regards these issues, these dilemmas at the frontier? If you accept what I have argued, the initial answer will be that any predictions can properly be identified only if the frame of reference is that of bioethics. Sadly, this is not the end of the matter. We then have to confront what is perhaps *the* emerging dilemma. Granted the priority of bioethics, how should bioethics do its work? I say this because bioethics is still very much in its infancy. Some even deny that moral philosophy has anything of relevance to say. It has nothing to offer, it is said, in the resolution of particular problems, whether in medicine or any other field. These are matters of politics – whatever that may mean.

Others who do not take this position nonetheless survey the current state of bioethics and quite properly criticise the disappointing level of much of its discourse. Bioethics, they point out, is in danger of becoming a ritualised incantation of certain formulae, a 'quadrilateral equation' of autonomy, beneficence, non-maleficence and justice. Just apply these and the solution appears! Such a mechanistic approach is, of course, the very antithesis of philosophy. Yet its easy appeal, offering an escape from the hard work of analysis, is alluring.

The claim of bioethics to set the agenda, to condition responses and thereby shape our predictions, depends therefore on its becoming better philosophy. Only then can it properly take on its role in explaining and facilitating understanding of what is at stake.

## THE STRUGGLE FOR THE HEART OF BIOETHICS

The shape of the future, if it is to be conditioned by bioethics, will depend first of all on the outcome of an old struggle recently given new life. Currently a series of assaults is being made on the liberal tradition which has dominated recent thinking in bioethics, the tradition which has stressed the primacy of respect for the individual. It is said by some that a preoccupation with individualism, an over-assertion of individual rights, has actually contributed to the production of dilemmas; respect for the individual should not provide the basis for future thinking and analysis. Instead the future of medicine, and hence our predictions, should be conditioned by a regeneration of the notion of the common good. The individual should be seen as someone having interests, or even duties, which are to be derived from – and in case of conflict reconciled by reference to – the common good.

The implications of such a shift in the frame of reference is obvious if we look, for example, at resource allocation or genetics. Unbridled individualism must give way, it is said, to the general good. We cannot, and a person has no right to expect that we should, allocate scarce resources to purchase expensive care for the mentally handicapped when they will never become able to fend for themselves. Nor has a person the right to expect geriatric care if it means that the young and productive receive less than optimum medical attention. The same arguments can be parleyed in the case of genetics. If a person has a heritable disease, or if information is available in the form of a genetic register, this should be made known. A person has no right to keep secret information which may materially affect fellow citizens.

For my part, having become a convert, I remain an adherent of a rights-based approach. Thus I would insist that my predictions be conditioned by a bioethics committed to rights. It is not that I am entirely convinced by rights-based arguments; I would be the first to point to severe foundational problems. It is rather that I recognise that most of the issues before us go beyond philosophy to social policy: they are not just matters for reflection but issues for decision and action. Given this, I am reluctant to see social policy shaped by reference to the common good. As a philosophical approach it has even greater weaknesses than a rights-based approach. As a basis for social policy, despite its apparently liberating quality, I am conscious of the effect recourse to it has traditionally had on the oppressed and the weak. I prefer therefore a bioethics of rights even if I recognise that my preference is shaped as much by the power of the rhetoric of rights as by argument.

## A RIGHTS-BASED APPROACH

Rights-based arguments do not, of course, immediately bring home the bacon. You still have to determine how to shape a theory of rights to accommodate the fact that all rights, or the rights of all, cannot be satisfied. We are forced then to ask what fundamental right it is, in the context, for example, of resource allocation, that we are urging. In my view, the answer is the right to equality. This, in turn, inevitably takes us into complex issues relating to what equality is concerned with and what must be equalised, questions which have recently been asked with renewed vigour by, for example, the distinguished philosopher Professor Amartya Sen. For my part, equality must be understood to mean equality of opportunity to have equal access to society's goods and services, preferring first, as part of the process of achieving such equality, the needs of the dispossessed, the poor and the vulnerable.

Rights-based arguments, when applied to the great medical issues of the day, are not free from problems. This can readily be illustrated by reminding you of two examples within the area of reproductive medicine which I have already touched on. The first concerns what can be described as maternal–fetal conflicts, that is, circumstances in which the conduct of a mother threatens to put at risk the fetus she is carrying. Here the question arises whether it is ever justifiable, and if so when, to place limits on the conduct of the mother for the sake of the fetus. At its most problematic, what is asked is whether a mother can be required to undergo some sort of medical procedure, say a Caesarian section, if this is the only way of protecting the continued existence of the fetus, when, for instance, the mother objects to such surgery on religious or other grounds.

The second example concerns the competing claims which can arise out of surrogate motherhood. In such an arrangement a woman agrees to bear a child for others. Problems arise when, for instance, the woman having borne the child refuses to give it up as arranged, or when the commissioning parents seek to reject the child on discovering that it is handicapped. Obviously the mere assertion of rights is not enough in either of these situations. The right being asserted must be carefully identified and some intellectual mechanism developed for preferring one party's rights to another's.

Rights-based arguments have other limitations. Some of the problems of bioethics are initially conceptual in nature. Whether an embryo is or is not a person, whether a family is to be defined by reference to biological, social or other criteria, whether a mother is the person whose genes a child carries – these are conceptual issues. Questions of the allocations of rights are subordinate; an embryo, for example, has rights if it is to be regarded as a person. This is not to deny that conceptual problems require moral analysis. Who is a mother turns on more than a lexicon's range of meaning; it calls for a response which arbitrates between various claimants based *inter alia* on moral principles. But moral analysis centred on a rights-based approach will not provide an answer. This can only come from examination of our ideas about the role of motherhood itself.

My conclusion so far is, however, that the discourse of bioethics should be rights-based, despite any theoretical shortcomings. Predictions will reflect this. If I say that I predict that embryo research will be condemned in Germany, what I mean is that the discourse will be ethical rather than scientific and will have recourse to rights-based arguments.

## PUBLIC POLICY MAKING

Bioethics is not, however, just discourse. It touches on, and seeks to respond to, the world of medicine and medical science. This is where some philosophers grow uneasy. For, in so far as bioethics offers a means of taking a view about medicine, it is a short but inevitable step to ask what should be done in the light of that view. Whether the answer is nothing or something, the fact cannot be escaped that we have moved from the realm of reflection to the realm of policy formulation.

How do we take account of this in our predictions? The first step is fairly easy to make, though hard to analyse. It can safely be asserted that responses to the issues I have touched on will be translated into some form of public policy. Unhappily, with rare exceptions, the criteria which dictate this process of translation are obscure and under-analysed. Witness the Warnock Report which, having decided that surrogacy was a bad thing, leapt without more ado to the view that it should not only be condemned but banned by law, and the criminal law at that. The body of knowledge concerned with the relationship between law and morals was sadly ignored, as was the opportunity to add to it by developing theory further and applying it to a novel and difficult set of facts.

I have argued elsewhere that the best means of exploring the criteria for public policy formulation would be a National Bioethics Commission. In such a forum these complicated issues could receive concentrated attention as a preliminary to developing specific public policy recommendations on any particular matter. The recent creation of the Nuffield Council on Bioethics may offer just this sort of opportunity. So let me turn to public policy. In general terms, in relation to any specific issue, it may express itself firstly in a decision that nothing should or need be done by way of public action; deciding how to respond to a particular problem may be recognised as being properly a matter for individual conscience and nothing more. Secondly, public policy may express itself in exhortations to observe the views of bioethics. These may take the form of guidelines or a circular issued by a Government or non-Government body, the former to assert and maintain its power, the latter to avoid real action. Thirdly, of course, public policy may take the form of law.

## PUBLIC POLICY IN THE FORM OF LAW

Let us assume that the discourse of bioethics will be followed by the formulation of public policy. Law is, perhaps, the most significant means of making public policy. Can we therefore predict the circumstances in which there will be resort to law?

Despite determined efforts to keep the law out of the world of medicine, I take the view that the conclusion will increasingly be drawn that law is the most appropriate mechanism for making public policy in bioethics. As a consequence law will become more and more involved in medicine. The pressure for legislation to deal with issues as they occur may be resisted, but it seems to me inevitable that the courts will be invited, urged or required to step in. Beset by problems which are immensely difficult, going to the heart of what we want for ourselves and for others, and faced by public institutions which are reluctant to act, those with something to gain or lose will take their claim to the courts, the one institution which, once asked, cannot refuse to supply a response. Of course, when the courts do step in, the subtle and difficult question of whether the issue really does call for legal regulation becomes moot. The court is stuck with the problem and must make a decision. Public policy there will be and it will be law.

## WHAT SORT OF LAW

Let me pursue the theme of public policy through law a little further. What predictions do I make, and how will the law respond? The most obvious form of law-making is legislation. Can we therefore expect legislation on the great issues of bioethics, the response of medicine to an ageing population, reproductive medicine, human genetics, resource allocation, access to and control of information, and so on? In some of these areas, we in Britain already have legislation. There has been a flurry of legislative activity over the past few years producing, for example, the Surrogacy Arrangements Act 1985, the Human Fertilisation and Embryology Act 1990, the Data Protection Act 1984, the Access to Medical Reports Act 1988 and the Access to Health Records Act 1990.

These, however, are either a product of the Warnock Report or of the need to respond to European initiatives such as the Council of Europe's Data Protection Convention. All could be said to have been grudgingly enacted. The Government avoided any legislative response to the Warnock Report for five years. It issued a Consultation Paper in 1986 and then a White Paper in 1987, and still the problems raised by the report refused to go away!

The reason for the general reluctance to legislate in these areas is not hard to find. None of the issues has any obvious party political content that will bring electoral gain to the government in power. All are troublesome and divisive. Compare, if you will, the ready willingness to pass the Animals (Scientific Procedures) Act 1986 regulating in fine detail the conduct of research on animals with the total absence of legislation relating to research on humans. Each issue is a potential mine-field. What government, for example, would wish to legislate on such a hugely difficult area as the selective treatment, or non-treatment, of severely handicapped new-born babies? There are simply no votes in proposing a bill which has as its object consigning some babies to death, even though the policy of doing so has wide support. The same is true when it comes to the sterilisation of vulnerable mentally-handicapped women. No government wishes to risk being accused of callousness, far less of eugenics, by statutorily endorsing such interventions. The practice goes on, apparently with widespread approval, but it is deemed better that it goes on without the involvement of government. Issues of bioethics are seen as calling for 'under the counter deals' for which no one is seen to be responsible.

The consequence is that calls for legislation will be as resisted in the future as they have been in the past. It does not require a political analyst to see that there are no votes in legislating to limit access to kidney dialysis to the under 55s, even though government may be happy to connive at it as a cost-saving technique. This is not to say that legislation is unnecessary or undesirable. A framework of law to respond, for example, to the issues of human genetics is urgently needed. But even if government could be persuaded to act, there would still be strident opposition from sections of the medical–scientific community who argue that legislation would stultify progress or would, by responding to current knowledge, set down on tablets of stone what is merely a stage in the process of understanding.

Those familiar with policy making can, of course, argue that such a view, applied generally, is a recipe for permanent inaction. It could also be pointed out that such critics do not reach for this argument when themselves urging legislation in other areas less close to home, whether it be airline safety or mental health. Such an attitude betrays a lack of awareness of what law is and what it can do. Legislative frameworks can be constructed which in no way inhibit development, although they may regulate or set the limits within which developments take place. An example, albeit of an international declaration rather than domestic legislation, is the Helsinki Declaration which outlines the circumstances in which research on human subjects may be carried out. All accept its premises, that research should be facilitated and at the same time the rights of research subjects should be safeguarded. None finds it an unwarranted intrusion standing in the way of future developments.

Legislation will continue, at best, to be slow to appear. Furthermore, attempts to meet the need by quasi-law (circulars, guidelines and suchlike) will be seen for what they are: helpful advice but ultimately a way of saying something without intending it to be taken too seriously. They will simply fail to meet the demand for public policy expressed as law. The courts will be drawn in, as I have suggested. This is already happening at an ever-increasing pace. Judicial decisions on issues of bioethics were rare only a decade ago; now it is impossible to open a law report or even a newspaper without reading of another case in which a court has had to leap where others have feared to tread. We have seen attempts to make a fetus a ward of court, and thereby endow it with legal personality. We have seen the Court of Appeal warning that it will not resist for ever the temptation to adjudicate on issues of resource allocation previously considered to be political and non-justiciable. We have seen courts agonising over the complexities of consent to treatment, non-consensual sterilisation, and non-treatment of a hopelessly disabled baby. The European Court of Human Rights on appeal from the UK has ruled on the civic status of a transsexual. Courts in other jurisdictions have wrestled with issues unimaginable only a few years ago, such as who should be regarded as the mother of a child born as a consequence of a surrogacy contract and who should be entitled to embryos frozen and stored when a husband and wife separate. This last question raises in turn the profoundly difficult question of the legal status of an embryo *ex utero*.

## THE RESPONSE OF THE ENGLISH COURTS

How have the English courts responded to their fast growing, and perhaps unwelcome, involvement in issues of bioethics? The first thing that can be said is that they have not approached problems on the basis of rights. I argued earlier for a rights-based approach. But the language of rights traditionally makes English judges nervous. They are unfamiliar with dealing in such currency; it is deemed foreign, European or American, associated with written constitutions and out of place in English law. Instead the courts have tended to be guided by perceptions of the general good, reflecting the notion of the utilitarian calculus which historically has pervaded the law relating to persons, if not property. Far from embracing the language of rights, the courts have barely even embraced the discourse of bioethics; their natural conservatism has meant they have looked backwards in prescribing for the future. Given also their traditional deference to medicine, a 'brother profession', the courts have by and large endorsed a medical view of the general good, deferring to the doctors' views of what would be best in any given case. To this extent the courts have adopted the discourse of medicine as the discourse appropriate to the analysis and resolution of issues of modern medicine.

Furthermore, the Courts, whatever their analytical inclinations, have often found themselves conceptually hamstrung. The concept of marriage, for example, is among other things a legal construct. Vast networks of arrangements have been made on the assumption that this legal construct has some certainty of meaning. If its meaning were altered, such arrangements could be put at risk with consequent social cost. This is a typical dilemma for a court when asked to develop or reshape some established notion. Aware of this and lacking the creative skills or courage to develop the law without undermining what has gone before, it is no surprise that the courts have retreated into the position of maintaining the *status quo*. It is for Parliament to make radical changes in the law, they say; they are only interstitial legislators. A marriage is reaffirmed as the union of a man and a woman for life. Homosexual unions cannot qualify as marriages. Transsexuals who are chromosomally men remain men whatever their psychological state, so any union between such a transsexual and a man cannot be a marriage. The logic is clear but somehow you may think that a court could do better.

Marriage is just one area in which new circumstances call for conceptual creativity. Another, fascinating example could arise over the legal status of an embryo now that the storage and use of embryos for research is allowed by law. Circumstances could arise, despite the best efforts of the legislators to avoid them, in which a court may have to decide just what an embryo is in law. Courts in other jurisdictions have already been confronted with this question and found that the conceptual tools available to the law leave something to be desired. The law knows only property or persons. If an embryo *ex utero* is a person, then clearly it enjoys the protection due to persons; it cannot be, for example, sold or destroyed. If, on the other hand, it is property, an embryo is on a par with an umbrella or a shoe, with all that follows from that. As well as being intuitively offensive, such a status is not in keeping with notions of the special status of the embryo. Yet in the absence of creative inspiration a court has only these two cards to play.

## THE COURTS IN THE FUTURE

Albeit falteringly, the English courts are undoubtedly coming to terms with the discourse of bioethics. The moral complexity of the issues has led even them to realise that not everything can be left to the doctors' judgement. So far they have largely eschewed the discourse of rights. Forced in the future to be in the forefront in responding to issues of bioethics, will they take the route of rights? My answer is unequivocally that they will, sooner or later. The single most important determinant in this process will be the gradual Europeanisation of English law. In particular, the European Convention of Human Rights will take on increasing significance, perhaps even being echoed by a parallel Bill of Rights in the United Kingdom.

Concurrently, medical law as it deals with issues of bioethics will come more and more to be recognised as an aspect of human rights law, a contextual application of the overarching principles of human rights. The legal culture which will set the mode of discourse and condition the response of the courts will become the culture of rights. This is my prediction: a prediction of a bioethics of rights with public policy in the form of law expressed in the form of rights.

Let me apply this conclusion for a moment to one of the great issues of bioethics which I have identified, the dilemma of an ageing population. It will mean that attempts through law or administrative action formally to curb the full rights of citizenship of the elderly, for instance by denying access to medical care which could be life-enhancing, will be seen as a denial of rights and fail. It will also mean that the right of the elderly who are ill to determine their treatment and, therefore, to end it will be recognised by the courts. The right to die with dignity will be reinforced as regards the comatose or those in a persistent vegetative state by accepting that nutrition and hydration, whether artificial or not, need not be delivered. The right to get help to die, perhaps unhelpfully called voluntary euthanasia, will be accepted. Whether it will be achieved through the historic compromise arrived at in Holland, where the State Prosecutor has indicated he will not prosecute doctors who help patients to die provided they follow the procedures agreed between the medical profession and the courts, or whether it will need legislation will be a test of the English law's maturity. The right to determine care and treatment after the onset of incompetence will be recognised in law by accepting the validity of what have come to be called Living Wills or Advance Directives, documents setting out in advance a person's wishes as regards medical care, especially life-sustaining treatment.

But I am descending to detail and problem-spotting, something I said I would not do. It is better I conclude by saying that I have put before you the manner in which I approach the issue of medicine at the frontier. I trust it serves to provoke thought on what for all of us are immensely challenging questions.

## FURTHER READING

Age Concern, *The living will: Consent to treatment at the end of life*, London: Edward Arnold, 1988.

Glover, J., *Fertility and the family*, London: Fourth Estate, 1989.

Mason, K., and McCall Smith, A., *Law and medical ethics*, 3rd edn., London: Butterworth, 1991.

Kennedy, Ian, *The unmasking of medicine*, London: Paladin, 1983.

Kennedy, Ian, *Treat me right*, Oxford: Oxford University Press, 1991.

President's Commission for the Study of Ethical Problems in Medicine and Biomedical and Behavioral Research, *Splicing Life: A report on the social and ethical issues of genetic engineering with human beings*, Washington DC: US Government Printing Office, 1982.

President's Commission for the Study of Ethical Problems in Medicine and Biomedical and Behavioral Research, *Screening and counselling for genetic conditions*, Washington DC: US Government Printing Office, 1983.

Thompson, J., *The realm of rights*, Cambridge, Massachusetts: Harvard University Press, 1991.

Weale, A. (ed.), *Cost and choice in health care: The ethical dimension*, London: King Edward's Hospital Fund, 1988.

# 6

## Divine providence in late antiquity

*AVERIL CAMERON*

Monotheistic religions have a vested interest in asserting the omniscience of God: man is granted only partial knowledge, and on God's terms. God is presumed to have a plan for the world and for humanity. But if God is good, why does he permit evil and suffering? What kind of scheme has he ordained for the world, and what place is there in it for human free will?

All these questions were debated during late antiquity, when the measures taken by the Emperor Constantine (AD 306–37) licensed the Christian church to develop as a public institution with imperial support, and permitted the eventual christianisation of the Roman empire. The doctrine of divine providence was not only integral to the Christian view of the world and man's place in it; it also explained the role of the church and justified its authority. Christian teachers therefore fought hard to preserve it in the face of fatalism on the one hand and total freewill on the other.

At the end of the period, the rise of Islam brought an even stronger emphasis on the will of God and the authority of scripture. For both religions, God's providential scheme, his plan for the future of the world, was expressed in a holy book. Not only that: both Christianity and Islam drew on and incorporated the sacred writings of the Jews. Thus each of the three great monotheistic religions sought to interpret divine providence in its own way and to its own advantage, on the basis of sacred books. But late antiquity is particularly characterised by the establishment of the authority of the Christian church and by the christianisation of the Roman empire. This chapter therefore focuses mainly on the efforts made by the church to win adherence for the Christian scheme of providence, and thus its control of ways of prediction, in the face of the stubborn attachment of contemporaries to fatalism, superstition, or secular explanation – what we might call rationalism.

## GOD'S PLAN, MAN'S WILL

As it was presented in this period, the very process of christianisation implied the acceptance of the scheme of divine providence and the authority of the church. But it was not easy to uproot existing patterns of thought. Among the miracles told in the early seventh century AD of the saints Cyrus and John at their shrine at Menouthis near Alexandria, one concerns an important individual called Nemesion, who made the mistake, from the Christian point of view, of trusting in fate. When he developed an eye disease (the speciality of these two medical saints), he spent a fortune on doctors, naturally to no avail. Only later did he realise, or so the story claims, that his disease was in fact caused by his mistaken fatalism. Doctors were unlikely to be of use – the only cure was true belief in the efficacy of Christianity and the healing saints.

Such improving tales abound from late antiquity. They are meant to point the moral lesson that man cannot rely either on his own foolish beliefs or on secular learning – only the church and its teachings can save him. A similar anecdote concerns a teacher of medicine who boasted that he had been baptised a Christian against his will, so as to comply with the law, and who liked to poke fun at the healing saints. Naturally he soon fell ill himself; the saints appeared to him in a dream to effect a cure, not losing the opportunity to point out that scientific medicine was no use in comparison with the power of God. The miracle stories would have us believe that both this man and Nemesion realised their error. Indeed, a dire fate was in store for those who failed to see the point: one such, an inveterate pagan who refused to admit the power of the saints, was overcome by demonic seizures and promptly died.

The church was evidently up against it. The Fathers had been writing on divine providence since the early Christian period, and from late antiquity a whole mass of writing survives, from more or less popular saint's lives and works of Christian instruction to sophisticated treatises aimed at rebutting belief in fatalism, astrology and the like in favour of a Christian doctrine of providence based on scripture and ecclesiastical authority. The argument often takes the form – especially in the later part of the period – of a stark opposition between supposed rationalism and true faith, with the latter always in the winning role. The church's claim to religious authority, backed up on many practical as well as doctrinal issues by the rulings of church councils, carried fundamental implications about the interpretation of the natural world, which the Christian texts held to have been planned and ruled according to God's intentions, according to a scheme which they called the 'economy of salvation'.

Within such a scheme God surely had foreknowledge. Yet a sufficient degree of freewill for men also had to be accommodated if they were still to be regarded as moral beings. It was a dilemma which called forth a range of subtle attempts to square the circle. But the appropriation of true knowledge by the church in contrast to what it condemned as worldly wisdom also had scriptural foundations. Some holy men took the saying that man's wisdom is foolishness with God so literally that they pretended to be fools themselves. One such was the Syrian, Symeon, who embarked on this life after twenty-odd years of austerity in the desert. In his efforts to shock, he went in for such exploits as bathing naked in the women's bath, or showing off his virtue by dancing with prostitutes. In this way he sought to demonstrate man's inability to understand God's ways; he, the fool, the channel for God's intervention, possessed the power of foretelling future events.

St Paul had written that 'if any man among you seemeth to be wise in this world, let him become a fool' (I Corinthians, 3, 18). But the piquancy of this and similar sayings, and of the whole posture of holy folly, was precisely that the fool was not foolish at all – on the contrary, he, the object of ridicule, misunderstanding and abuse, was the one with the true knowledge and true rationality. In a less extreme form we find this theme repeated throughout patristic literature. In late fourth-century Pontos, for instance, St Gregory of Nyssa claimed that the allegedly simple faith of his sister, the Christian lady Macrina, brought up at home and educated only on the Psalms, converted their brilliant and intellectual brother, St Basil, fresh from influence of the clever teachers of Athens, to the true philosophy of Christian holiness.

Late antique man was pulled in opposite directions, between the natural world and the spiritual, between rationality and faith, between scientific prediction and unforeseen miraculous intervention, between the stars on the one hand and God on the other, and between freedom and predestination. Nor was it a matter only of popular superstition and popular fatalism: an important part of the Christian argument was directed against pagan philosophy, and neo-Platonists were in turn preoccupied with the themes of fate, causation and prediction. Whatever the church liked to think, the competition was all too real, and the edges of Christian faith remained obstinately blurred.

In effect, the idea of Christian providence constituted a totalising explanation, a kind of theory of everything. It embraced the idea of a divine plan which began with Creation, progressed through the Incarnation and culminated with the Second Coming and the Day of Judgement, and in which all history was subsumed. In the working out of this providential scheme, Christianity took from Judaism the ideas of God's promise to man, and his plan for the world, and indeed Paul says in the Epistle to the Romans that the new dispensation of Christ transformed God's promise made to Israel but did not nullify it. But that promise had been expressed in the Jewish scriptures. Much of the discussion therefore centred on the exact nature of that promise and the interpretation of the prophetic texts on which Christian claims also rested.

Taken to its logical conclusion, christianisation implied a shift in mentality, a new cosmic image or world model. But the sheer volume of argument going on in this period over the fundamental issues of providence, freewill and divine law leaves one with the impression that things were not nearly so straightforward. I see late antiquity rather as a time of intense competition between different systems of knowledge, each trying in different ways to resolve the basic questions of human freedom and divine purpose, and to locate man in an explicable world and an explicable universe. This led, of course, to contradiction, for the mental repertoire of any one individual can and usually does contain at one and the same time material from several different and often quite contradictory and logically inconsistent systems of thought. Indeed that may be the characteristic way in which human beings cope with the contradictions of life. Let us see therefore how these potential dissonances manifested themselves in late antiquity and what strategies were adopted for dealing with them.

## AN ORDERED WORLD

First, belief in the possibility of prediction assumes that the physical world is controllable, whether by appeal to religious authority or by scientific truth. Part of the debate in late antiquity therefore centred on the nature of the material world and the laws which were held to govern it.

It was not in fact obvious to everyone in late antiquity that the history of the world was looked after by a beneficent providence. A substantial group, particularly the Manichaeans, to whom the young Augustine belonged, held that matter itself was evil, and saw the world as such a miserable place that it could not have been created by a good and just God. Instead they posited two Gods, one of whom, the Creator, was himself connected with evil matter. In such a world man had little room for freewill and determinism reigned. Orthodox Christian writers spent much effort in arguing against such ideas and asserting the conception of one good God who had himself created the world according to a scheme of beneficent providence, and learned treatises of ten books or so *On Providence* were nothing unusual. Nor was the matter easily settled: fatalistic dualism continued for centuries to occupy a central place in the catalogues of heresies repeated and condemned by generations of orthodox polemicists.

In contrast, the creation of the world by a good creator according to divine providence early became an essential component in the orthodox Christian armoury. While it was a difficult task to accommodate belief in freewill within the conception of an overall divine plan, at least it could be agreed that the notion of an evil world had to be rejected. But a different challenge presented itself from another direction, which if accepted would completely undermine the ideas both of a divine creator and of a beneficent providence, namely that the world had not been created at all.

## CREATION

Christian writers had first to be clear about what they thought about creation themselves. From the Platonising Basil and the sensible Augustine to the literalist Theodore of Mopsuestia, they therefore struggled over the right way to interpret the creation stories in the book of Genesis, and most of the major Fathers wrote commentaries or treatises on the subject. Both the level of literalism with which the Genesis account was to be interpreted, and the apparent conflict between Genesis and Ptolemaic cosmogony were fought over again and again. Ambrose and Basil were only two of those who composed, in Latin and Greek respectively, 'hexaemera', commentaries on the account of the six days of creation in the book of Genesis; another was Augustine, who wrote two works, *On Genesis against the Manichaeans* and *On the literal interpretation of Genesis*. However they might argue individually, when taken as a group such treatises represent a series of attempts to save the text, to transform the stories into an intellectually defensible theory of matter and divine purpose. Whether the creation story was to read literally or allegorically, it was part of scripture and it had to be kept.

Plato's *Timaeus* also offered a conveniently creationist account which could be turned to advantage by Christians, many of whom like Augustine were deeply influenced by Platonic teaching. But even neo-Platonists were disposed to read the *Timaeus* non-literally, and some, like Proclus, affirmed the doctrine that the world had no beginning. On a more straightforward level things should have seemed easier. The seventh-century miracle story already cited turns to Genesis in the course of its attack on belief in the stars, to show that the heavenly bodies are themselves under God's providence: 'God said, "Let there be lights in the vault of heaven to divide day from night, and let them indicate festivals, days and years" '. But the argument in fact derives from long-established discussion on the nature of the heavenly bodies. In their zeal to argue against astrology, many Christian writers went out of their way to argue that the heavenly bodies were mere matter, with no possible influence over human lives, and the belief that they were moved by intelligent minds was the subject of imperial condemnation under Justinian. In response, the neo-Platonist Simplicius complained in his commentary on Aristotle's *De Caelo* against his Christian rivals that with this kind of argument they dishonoured the true glory of the heavens.

Yet this was not, as it might seem, a straightforward argument between religion on the one hand and rationalism on the other. Certainly Simplicius defended the theories of Aristotle, but his own neo-Platonist approach was equally religious. The spectrum of opinion was in fact very wide. Belief in astrology was evidently widespread at a popular level, and it was even still possible in the sixth century to write a serious astrological treatise and to claim against the tendency to ascribe disasters to divine providence that they were caused by natural phenomena.

### THE CHRISTIAN COSMOS

Nevertheless, in the sixth century the issue of creation presented itself as a direct conflict between neo-Platonism, Aristotelian science and Christian fundamentalism. The nature of matter was something which would continue to bother Christians into the early Islamic period.

In the mid-sixth century, a certain Cosmas produced a work known as the *Christian Topography*. Often derided for its simple-mindedness, it argues for a literal reading of Genesis and a flat earth made on the model of the ark of the covenant, with heaven above it. There are many absurdities in the work to modern taste, but Cosmas's basic ideas were much more commonly held than this might suggest, and the work was widely used and very widely copied. Though as usual we have manuscripts only from a later period, it seems that it was

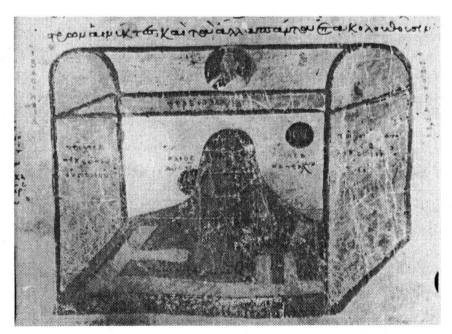

*Figure 1* The world as depicted in the *Christian Topography* of Cosmas Indicopleustes. Heaven and earth are divided by the *stereoma*, the firmament. Eleventh-century illustration

illustrated with diagrams and pictures showing the structure of the world and the cosmos according to the author. Despite its apparent simple-mindedness, the *Christian Topography* engages with contemporary philosophical ideas about creation and the nature of the material world, and is therefore of some importance in understanding late antique cosmogonies and ideas about the place of creation in God's providential plan.

Cosmas opens with an introduction to the work as a whole, and a justification for the enterprise: 'some who are supposedly Christian pay no attention to the scriptures, and disdain them like secular philosophers, misled by solar and lunar eclipses into understanding the world as round'. His own work is an ambitious exposition of the whole of Christian knowledge. Book I explains why Christianity is incompatible with pagan philosophy; Book II outlines the scriptural teaching about the world; Book III demonstrates the accuracy of Old Testament prophecy, and Book IV develops the proposition that the world is flat. Finally, Book V shows how the flatness of the earth relates to the whole providential scheme of God from creation to the life to come, which, as the author says, 'Jesus Christ will enter first, after his bodily resurrection, and after him the just'.

Cosmas defines three groups of people – 'true Christians', pagans, and 'false Christians' – among whom the last (those who allegedly try to combine Christianity with Greek scientific and philosophical ideas) are his real target.

Naturally not all Christians were such fundamentalists. One in particular, John Philoponus, came from a milieu in sixth-century Alexandria which sought to reconcile Aristotelianism with Christianity. Though Philoponus himself argued against Aristotle, he also (like Augustine) argued against the literal interpretation of Genesis. In his early work, *On the eternity of the world against Proclus*, Philoponus had supported belief in creation against the neo-Platonist view that the world was eternal; a later work by him, *On the creation of the world*, offered a more enlightened reading of Mosaic cosmogony, attacked astrology and was probably in fact directed against Cosmas.

For the philosophically sophisticated, various compromises were possible. One was the neo-Platonist view that the material world in its present form had a beginning, but that matter itself was not created. Against this view, Philoponus argued that matter itself, not just the present world, also had a beginning. The anti-spherical cosmology, with its appeal to scriptural authority, also became widely accepted. The secular Greek historian Agathias describes the arguments which arose in Constantinople about Aristotle's theory of earthquakes after the earthquakes of AD 551 and 557; evidently many educated persons upheld Aristotle's explanation of earthquakes in natural terms, but Agathias himself preferred to put his faith in the divine will rather than in science.

Cosmas Indicopleustes was an Antiochene literalist. Yet whatever their source, his basic ideas came to prevail. Their authoritarian and hierarchical thrust chimed in with government attempts to impose religious orthodoxy; thus in the sixth century pagans were sought out, imprisoned and exiled if they could not be converted, while stronger measures were brought in against Jews. Yet it seems that paganism persisted. A show trial involving important personages (including a patriarch) was held in Constantinople in the late sixth century, pagan prophecies were in circulation and pagan practices were condemned as late as AD 691 by the Council in Trullo, while 'secret pagans' feature commonly in the collections of miracle stories and the question-and-answer literature of the sixth and seventh centuries as a standard sub-group among other recalcitrants such as heretics, Jews, Manichaeans and Saracens.

## THE FUTURE: THE REWARDS OF THE JUST

But even if it could be agreed that the world had a beginning, what of its end? It was not doubted that God's providence encompassed time: there would be a Second Coming, an end to the world and the resurrection of the just in heaven. Exactly when and how this would all happen was, however, a matter for judgement in the light of experience. Moreover, God's expected interventions in history on the side of the just did not always seem to happen as they should. When Christian Rome was sacked by Visigoths in AD 410, Augustine explained in the *City of God* that Christians were mistaken if they imagined that their faith necessarily implied blessedness in this world; rather, they were still being tested, and the kingdom of heaven was yet to come.

When things did apparently go right, on the other hand, much was naturally made of it. The destruction of the Jewish Temple provided a favourite theme; Cosmas, like many other Christian writers, used it as an apologetic argument in support of Christianity, which he claimed was diffused through the whole world. Together, the destruction of the Temple and the spread of Christianity were taken to justify the Christian interpretation of Old Testament prophecy.

At the other end of the scale, Christian writers also devoted much debate to the exact details of the final judgement, and particularly to the rewards which the just could expect. Much careful attention was given to how individuals would experience resurrection, and in particular to the nature of the resurrection body; would body as well as soul be resurrected, and would there still be male and female in heaven? John Philoponus wrote a tract in which he discussed the question of whether the material and corruptible body would be replaced by a superior and incorruptible body or merely refashioned. Apart from the body, what about the soul, was it too corruptible? This argument was the cause of bitter division in the sixth century and caused several individuals to have their books burned as heretical.

## CHRISTIAN TIME

The idea that God intervened for a good purpose in the physical world, in history, and therefore in time, and that in due course time itself would end, was fundamental to Christianity. It was, after all, a religion which placed its main emphasis on the Incarnation, an event in history in which God himself took on human flesh. The paradox whereby the creator became himself a creature thus provided the theme for many word-plays in Christian poems and homilies.

Yet belief in such a divine intervention in history also transformed the concept of time itself, and Augustine famously devotes an extensive discussion to time in the *Confessions*. Before God created the world, he argues, time did not exist; thus one cannot logically ask what God was doing before he made heaven and earth. The same theme was also taken up in Boethius's *Consolation of Philosophy*, and Augustine's own greatest work, the *City of God*, is in effect an extended meditation on providence and history. At a more mundane level, Christian world chronicles started with Adam and typically continued up to the writer's own day: history was to be read as a linear process, a working out through time of God's plan for the world. Time, too, was being ordered and reshaped.

## HOW TO PREDICT

In these and other ways Christian writers in late antiquity tried persistently to place man's role in the material world in a context of ordered time and space. I want to turn now to their attempts to control prediction as well.

People used all kinds of methods to find out the future. One was astrology. To judge from the frequency of attacks on it and of assertions that the heavenly bodies were mere matter, belief in the stars remained extremely popular; indeed the confrontation of the astrologer with the Christian holy man is a standard theme in saints' lives. Just such a debate with astrologers was allegedly held by St Symeon the Stylite the Younger, a pillar saint from the late sixth century with a considerable gift for prophecy (he could also foretell future emperors). The same questions were evidently addressed to astrologer and saint alike, and the latter was often put on his mettle to establish his credentials.

The virulence of the denunciations of astrology indeed suggests that attachment to it was felt to be a real threat to Christian control.

*Figure 2* Silver plaque showing St Symeon the Stylite, probably Symeon the Elder (*c.* AD 389–459) on his pillar at Qal'at Sem'an near Antioch. The serpent may refer to an incident in Symeon's life, or it may symbolise the Devil. Sixth century

Straightforward astrological prediction and resort to chance were repeatedly condemned in Christian works, and dice, magic and soothsayers of all kinds were formally attacked by the church. In practice they were evidently everywhere to be found. One of the stories told of Symeon the Fool turns the tables on them by recounting how Symeon (in his fool's guise) first befriended a soothsayer and amulet-maker, and then promised to make her an amulet which would prevent her ever being touched by the evil eye; when she accepted, he slyly gave her an amulet on which was written in Syriac 'May God render you ineffective'.

Then there was recourse to oracles. These were a major target for Eusebius of Caesarea, who wrote in the early fourth century when pagan oracular shrines still represented a very real threat. The scale on which the big oracular shrines worked in the imperial period has been vividly shown by recent research, and it was essential for the church to persuade people that they were mistaken in frequenting them. When Eusebius was writing, Christians themselves had until very recently been objects of persecution, and paganism was very much alive, as can be seen from Eusebius's eagerness to claim that when the persecution ended no protest was raised by the pagan gods. As for pagan oracles, where were the gods in their oracular shrines, he asks, and why did they not cry out in protest when the emperor's men came and took away their statues? For Eusebius, the 'true oracles', the reliable source of prediction, are the scriptures themselves.

Later in our period, healing shrines of saints, with their associated miracle stories, took over the role of pagan oracles, and adopted many of their cult practices, such as pilgrimage and incubation. The saints spoke in dreams just like the pagan gods who were their predecessors, and these dreams too, just like the earlier ones, usually needed specialised interpretation. Pagan priests in charge of oracular shrines would translate the divine messages into intelligible information; the redactors of miracle stories would write up the dreams described by those who consulted the saints, turning them into cautionary tales of the type that we have already encountered.

Even in the seventh century, it was still felt necessary to muster arguments against pagan oracles. Christians asked themselves how it could be that pagans, heretics and non-believers nevertheless had the power of prophecy, and received in answer a version of the familiar 'God has his own purposes' argument: such people did not prophesy in their own right, rather, when it suited him, God could choose to deliver signs even through non-believers.

A Christian could also open the scriptures at random for an answer to a particular question. Though the church was not happy about it, the practice was evidently widespread and receives a grudging approval in a seventh-century manual of instruction. Going to a holy man for information about the future was almost as common, and more acceptable in the church's eyes; late antique hagiography and other sources are full of such stories. St Benedict, for instance, is said to have predicted the fall of Rome in AD 546 to the Gothic king Totila from Monte Cassino. Pope Gregory the Great, who tells this story, is also a rich source of information about other holy men who could read thoughts and foretell the future; the prophecies of the dying were thought likely to be specially reliable.

But together with resort to such methods as soothsaying, throwing the dice, and so on, there also went the uneasy fear that the messages received might really come from a demon. A woman at the Golden Gate of Constantinople once threw a crowd into turmoil by announcing the end of the world; they panicked not so much through fear of the end, but because of the suspicion that she was controlled by the devil. To counter such false prophecy, as well as belief in blind fate, the Council of 691 condemned all charlatans – necromancers, fortune-tellers, genealogers, makers of amulets, 'those who pretend to be possessed by demons', and 'those who sell animal hair for charms'.

Contemporary texts give the impression of a virtually universal desire for certainty and foreknowledge, obtained by whatever means, in a context in which however much the church might try to persuade itself otherwise, a man or woman was as likely to go off to a soothsayer as to follow precept and go to church. Indeed, the impression given by some of the sources about Christian practice is one of considerable laxity to judge from the complaints about the bad behaviour of congregations, yawning, talking and showing a general lack of respect. The church hit back in this area by attempting to lay down those forms of prophecy and revelation to which it lent its authority. These consisted first and foremost of the teachings of the church itself as revealed in the ecumenical councils.

The problem was that the latter only covered a small part of what an ordinary person might want to know, as we can see from the many quite mundane issues raised in seventh-century question collections. The Fathers could also be consulted (the appeal to tradition), but there again such consultation needed to be filtered through the right channels, or the wrong texts might be cited. To facilitate matters, collections of approved quotations were drawn up, so that all that had to be done was to consult them on a given point. Such an appeal to authority became a *sine qua non* of theological argument, from the informal and popular questions and answers on all kinds of ordinary matters to the high-level debate at councils.

There were still other means of divining God's will – recognised signs through which it was held that God's providence was made known. These ranged from the Cross, the Gospels and the Eucharist to the signs in the Old Testament – the burning bush, the ark, the tablets of the law, the dew upon the fleece, manna, and so on, all of which were frequently cited by Christians as ways in which God had revealed his providence to men. The status of these signs was disputed against Hellenes, Jews and (later) Saracens, as well as within Christianity itself, but the various Christian groups all shared the same assumption that knowledge was indeed based on divine revelation, part of God's providence.

## THE BOOKS OF PROPHECY

There was thus through the period of late antiquity a steady, if slow, impetus towards a system of knowledge based on fundamentalism and authority. Not surprisingly, it was a system in which Old Testament prophecy occupied a special place.

In Book V of his *Christian Topography*, Cosmas Indicopleustes parades one by one the patriarchs who have foretold the coming of Christ from Adam to Moses, the figure and type of Christ, 'cosmographer of creation', who received the signs of the law and the burning bush. He then turns to David, Elijah, Isaiah, Jeremiah and the rest in a classic florilegium of Old Testament prophecies. Only after this does he move on to the New Testament, beginning with John the Baptist, the forerunner, and then the Gospels, the evangelists and St Paul. 'Why was the law given?', he asks, 'In order to reveal the coming of Christ'. It is an explicit assertion of Christian over Jewish claims. The whole reveals a perfectly coherent plan, culminating in

*Figure 3* The *Christian Topography* of Cosmas Indicopleustes:
the hierarchy of salvation, with Christ enthroned in glory above
the angels, the living and the dead. Ninth century illustration

the final judgement, of which 'no one knows the day, save only God'.
The thinking is not merely linear but also hierarchical: Christ will at
the last be seated in glory, with the ranks and hierarchy of angels and
men beneath him in order, and below them the dead who will be
raised.

134

This comprehensive systematisation and interweaving of the old and the new dispensations allowed room for no opposition. Only two conditions were possible: that of the present, in this world, and that of the future, revealed clearly by Christ and 'in shadows' by the prophets. These conditions are foreordained by God and exist in a temporal relation to each other. Pagans, false Christians, Manichaeans and Marcionites, all other heretics, Samaritans and Jews are condemned together.

It was an over-confident scenario. Though Cosmas makes a feeble attempt to explain why those benighted groups still existed in the sixth century and how they could expect to be punished, in general the untidiness of the real situation was ignored. Some of the catechetical questions concern the equally difficult problem of why the just die young and the unjust prosper. But a much more fundamental challenge to Christian ideas of the economy of salvation arose in the early seventh century when in 614 Jerusalem itself was sacked by the Persians, the True Cross taken off to Ctesiphon, and the Jews left briefly in charge of the holy city.

From the Christian point of view, God's role in history was truly in need of defence. Both Jerusalem and the Cross were as it happened dramatically recovered by the Emperor Heraclius in AD 630, but by another dramatic reversal Jerusalem was lost again to the Arabs in AD 638. Ironically as it seems in retrospect, it took some time for the church to realise that the Arabs posed a religious and ideological threat to Christianity as great as or even greater than the Jews. In time, the status of divine writ and the holy book as the source of prophecy would be fought over by Christians, Jews and Muslims alike, but during the seventh century, even after the Arab conquest, the spate of Christian argument about prophecy still addressed itself mainly to the Jews. Tests were marshalled to prove that Moses did indeed prophesy the coming of Christ and that Jesus was indeed meant in the passages prophesying the Messiah.

The same repertoire as that used by Cosmas comes out again in lists of citations, standard Old Testament texts forming a whole battery of argument. One is not surprised to find that when Christian/

Islamic polemic did begin in earnest in the eighth century, such was the weight of accumulated argument about prophecy and the status of prophets that the same texts were deployed in very similar ways.

## SAVING THE FAITH: WAS FREE WILL POSSIBLE?

Prophecy apart, without free will man would have no moral responsibility and there could be no last judgement. In the armoury of writing on the subjects of providence and determinism, fate, allegedly the favoured doctrine of pagans, was a relatively easy target for Christian writers. The trouble really arose when they tried to reconcile God's providence with human freewill. Where did the one end and the other begin?

Certainly, a thoroughgoing belief in fate left no room for freewill; in whatever form it was conceived, or by whom, fatalism could easily be shown to reduce man to utter dependence, take away moral freedom and interfere with God's judgemental role. The theme was and is a classic one, and as many scholars have shown, one Christian writer after another took up and reworked traditional stoic arguments on the subject. But it was far more than just the repetition of a traditional theme. The need to combat belief in fate and get the balance right between free will and providence constantly presented itself in new guises to Christians of this period. To give only a few examples, Paulinus of Nola, writing in the early fifth century to a pagan friend, listed at length the arguments for divine providence over fate, while in the late fourth century St Ephraem the Syrian argued against Marcionite and Manichaean dualism and St John Chrysostom wrote six discourses on fate. St Basil's commentary on Genesis typically took the opportunity to argue against astrology. There were dangers at both ends of the spectrum: in AD 529 in the west the Council of Orange, even while supporting Augustine against a Pelagian over-emphasis on freewill, felt it necessary, nonetheless, to condemn predestination.

From the end of our period we have two Greek treatises dealing with the theme of the predestined terms of life, that is, a particular manifestation of fatalism whereby the manner and date of everyone's death was held to be foreordained. They were written, nearly a century apart, by a classicising historian and a patriarch of Constantinople respectively. The first uses all the armoury of rhetoric, yet puts the argument completely in scriptural terms: two disputants address each other in set speeches, both bringing up strings of biblical proof tests. Then there is a formal adjudication.

The verdict is 'Yes and No'. The idea of complete predestination, the judge says, is 'a Greek concept, a *proprium* of an autocratic Destiny'. God has indeed decided that human life is finite, and part of the system of divine providence, but he has also given man free will, so that he can choose between good and evil. Man, not God, is responsible for sin, but finally the ways of God are mysterious; as Paul said, 'we see in a glass darkly', and knowledge does not lie with the so-called wise.

The later treatise derives its arguments, like the first, from scripture and from earlier Fathers, especially St Basil, who had himself suggested that an individual's end was determined by God. It is addressed formally to one of the high dignitaries in Constantinople, from which we may infer that there was widespread interest in the topic. In fact the same issue or something close to it is also discussed in a sixth-century historical work in relation to earthquake victims. The only notable person who died in a particular earthquake in the capital was, it is claimed, a certain Anatolius, a high official much hated for his rapaciousness and confiscations of property. The author puts the dilemma well: people naturally jumped to the conclusion that he had been punished for his sins, but then what of all the others who were just as bad as he was? It is socially and morally useful to retain the idea that wickedness will be punished, but in the end only God knows how and when this will be done.

Questioning the efficiency of divine retribution implies questioning both the notion of God's goodness and the belief in God's power, and late antiquity experienced many earthquakes and plagues – not to mention enemy attacks – which caused people to wonder about both. That God is both good and powerful, and that there must be a divine plan even if we cannot understand it, is the conclusion of our second treatise on the terms of life: the same answer as is given to many questions on the subject in seventh-century collections of questions and answers, and by one of the last major Greek historians, Procopius of Caesarea, when contemplating the reasons for the sack of Antioch by the Persians in AD 540.

Paul had grappled with the question of grace and predestination in the Epistle to the Romans. In late antiquity, the question was resolved in such a way that God's credentials were saved. Man was given responsibility for his own suffering and misfortune, through sin, yet somehow God had also revealed a beneficent plan for the history of the world. It was not so much an answer as an assertion of faith.

But a similar end could also be reached by different means. In Boethius's *Consolation of Philosophy* the Lady Philosophy appears to Boethius in prison with arguments meant to assuage his mental torment, which she first diagnoses as consisting in regret for his previous good fortune. Change, she says, is the very nature of Fortune as she turns her wheel, and Fortune rests her case on the mutability of human life: someone who has been so fortunate as Boethius cannot now complain.

Philosophy's lesson is that happiness must lie elsewhere. Rather than resting in external goods, which are prone to sudden change, true happiness must lie rather in the mind, and especially in the capacity of the wise man for detachment. The long poem which Boethius places in Book III derives both from Plato's *Timaeus* and from the commentary on the *Timaeus* by the neo-Platonist Proclus; it eloquently expresses the ideas of beneficent creation, divine providence, cosmic harmony and the longing of the soul for return to God. It was only to be expected that Philosophy would go on to discuss the other neo-Platonist topics of providence and fate, the former in the higher sphere, the latter in relation to cause and effect in the material world. Finally, the *Consolation* concludes with a famous discussion of free will and necessity.

Between this classic work and the treatises I have mentioned there are both similarities and contrasts. Whatever his personal religion

*Figure 4* The Lady Philosophy consoling Boethius. Fifteenth century illustration

may have been, Boethius does not draw as they do on the stock of scriptural citations to prove his points. Nor – though freedom is for him a moral responsibility – does he focus on individual suffering, emphasising rather the intellectual problem of reconciling free will with providence. His antecedents are philosophical; he draws on a dense tradition of commentary on, and development of, Plato and Aristotle. Like others before him, he arrives at a compromise

*Figure 5* The wheel of fortune. Fifteenth century illustration

solution, distinguishing between two sorts of necessity, of both of which God has foreknowledge. There is thus the kind of necessity inherent in the fact that all men must die, but there is also another category in which free will is involved. In the latter case, it is possible to change a particular instance but still not possible to evade providence, since providence has foreknowledge of both kinds of eventuality. 'God has foreknowledge and rests a spectator from on high of all things.' Hope and prayer are therefore not pointless, though in order to be effective they must be of the right kind – it is no use hoping or praying for mistaken objectives.

Finally, the work concludes, 'a great necessity is laid upon you, if you will be honest with yourself, a great necessity to be good, since you live in the sight of a judge who sees all things'. A transcendant God, a beneficent providence, *and* the possibility of free will were common to Christians, Aristotelians and neo-Platonists alike.

### WHY?

How the world was directed, whether there was a master plan, and if so how one could discover what it was, are questions which recurred again and again in late antiquity in all sorts of contexts, from the intellectual heights of philosophy to the daily lives of ordinary individuals. I think we should ask ourselves why this was so.

Was this, as we are often told, an age of spirituality, a more spiritual time in history, a time when people were more prone to look for answers in religious terms, and more prone to believe them than we are ourselves? Or was it rather, on a less favourable but equally common view, a period when the Roman empire suffered an unfortunate slide into credulity, superstition and irrationality, encroaching gloom or what you will – the beginning, in other words, of the Middle Ages?

Both are surely over-simplifications. Late antiquity was evidently a time of competing systems of thought, with a correspondingly high level of dispute and argument. Though this is the time when we can realistically talk of christianisation, or at least of an increasing amount of control exercised by the institutional church, there was still a whole range of opposing beliefs for the church to attack. The competition was a live one, and it was conducted on many fronts. All kinds of tactics were brought to bear besides argument alone – from carrots on the one hand to sticks on the other – yet still there was resistance.

The debate went on. Free will became a major topic under early Islam, in reaction to the greater emphasis placed in Islam on the will of God; it was also a theme earnestly reasserted by Christian controversialists writing in Greek, Syriac and Arabic. But we need look only as far as the preamble to the new law code issued by the Byzantine emperor Leo III in the same period. God, state and church are brought to bear together on the individual. The law is seen as God's will on earth and the emperor's duty is to carry it out. Constantine had already adopted the same *persona* in the fourth century, and he was not an emperor renowned for humanitarian legislation. Leo went a stage further. It was his law code, brought in in the name of Christianity and as the work of the reforming emperor of a Christian commonwealth, which endorsed the elaborate system of mutilation, nose-slitting, hand-lopping, tearing-out of tongues, and so on which has helped to give medieval Byzantium such a proverbially bad name.

Yet despite the undoubted thrust in the direction of authority and even coercion which is so apparent in the texts we have considered, it may still be best to conclude by deconstructing them, and by suggesting, however unfashionably, that these very texts are evidence that late antiquity was after all not so very different from the present day. Individuals harboured within themselves the same bundles of incompatible ideas; not everyone was either superstitious and credulous or given to secret paganism. Aristotelian science and the arguments of highly intellectual philosophers like Philoponus or Simplicius no doubt hardly reached the average person (though there were street fights between pagan and Christian students in sixth-century Alexandria and we are told that the people of Constantinople could get very excited when a bishop was put on trial for paganism).

In our own society we have our soothsayers, our amulets and especially our faith in rational explanation, even if the latter has become a little dented of late. Western science and western rationalism have until recently provided a modern orthodoxy, and in a sense a modern fundamentalism, able to stand comparison in structural terms with the Christian orthodoxy of late antiquity. They are now challenged on several sides, by the emergence of other kinds of fundamentalism, by a post-modern attachment to cultural pluralism, and by an expansion of world horizons on such a scale as to render all types of prediction highly uncertain.

Late antiquity was characterised in contrast by the energetic dissemination, and in many cases the attempted enforcement, of a hierarchical Christian world-view claimed to provide answers to a wide range of human problems and a sure means of predicting the future. How completely it was accepted in practice remains, of course, debatable; but it was not a scheme which made for toleration or allowed for pluralism or dissent.

## FURTHER READING

Amand, D., *Fatalisme et liberté dans l'antiquité grecque*, Louvain: Bibliothèque de l'Université de Louvain, 1945. Reprinted Amsterdam: Hakkert, 1973.

Brown, Peter, *The world of late antiquity*, London: Thames and Hudson, 1971.

Dargon, G., 'Le saint, le savant, l'astrologue', in G. Dargon, *La romanité chrétienne en Orient*, London: Variorum, 1984, no. IV.

Dihle, A., 'Liberté et destin dans l'antiquité tardive', *Revue de théologie et de philosophie* 121 (1989), 129–47.

Sorabji, Richard (ed.), *Philoponus and the rejection of Aristotelian science*, London: Duckworth, 1987.

Watt, W. M., *Free will and predestination in early Islam*, London: Luzac, 1948.

# 7

## Buddhist prediction: how open is the future?

*RICHARD GOMBRICH*

The term prediction suggests a concern either with natural science or with eschatology. It suggests the grand scale: the prediction of eclipses, or the Last Judgement. To use the term for the little events of human life tends to sound, at least outside the laboratory, rather portentous.

Yet it is human, not cosmic, predictability that lies close to the heart of the Buddha's teaching. For the Buddha was concerned neither with natural science nor with eschatology. He was interested in the microcosm of the living individual, not the macrocosm; and his point of view was that of the doctor, not of the research scientist. If any kind of prediction concerned him, it was what we would call prognosis. Developed Buddhist dogmatics differentiated two uses of the term meaning 'world' (*loka*). The material universe was dubbed 'the world as receptacle' (*bhājana-loka*) and was of marginal interest; it was merely the stage on which the real drama of 'the world as living beings' (*sattva-loka*) was forever being enacted.

Buddhism is a soteriology, a way to salvation. Salvation is a property, or achievement, of individuals. In the Christian tradition there is also a concern for the fate of human society conceived as a whole, rather than merely as a sum or network of individuals, a concern reflected in the doctrines of the Second Coming and the Last Judgement. But Buddhism is more severely analytical: not only does it dissolve society into individuals; the individual in turn is dissolved into component parts and instants, a stream of events. In modern terminology, Buddhist doctrine is reductionist.

Like other religions of Indian origin, Buddhism draws no stark dividing line between human society and the rest of the animate world. Buddhists believe that living beings constantly die and are reborn according to their moral deserts in a many-tiered, hierarchically arranged receptacle world. Various classes of gods inhabit the spheres above us, and may visit us; human beings share this world with animals and spooks; and in hells down there, demons torture the wicked. Power, longevity and general well-being increase as one goes up the system, but no life is eternal – even gods die – and to that extent at least there is no perfection anywhere. Whereas life and hence happiness are inevitably finite, however, Buddhism sees the universe as infinite in space, beginningless in time, and of unlimited duration. There is no creator god and no final destruction. The Buddha argued that everything arises from a cause; so there can be no beginning to the world in either sense, as receptacle or as a community of living beings. The latter, dismayed by the suffering of reiterated rebirth, try to put an end to it by attaining nirvana; until all succeed – and it is not envisaged that that will ever happen – the receptacle world must in some form survive to accommodate them.

All this is in contrast to Christian belief. But perhaps the most important contrast between Buddhism and Christianity has to be made on another level. It is simply that the critical study of Buddhism is still in its infancy. In sheer volume, critical work on Christianity must outweigh that on Buddhism by something of the order of a thousand to one, if not more; and so great a disparity in quantity cannot but affect quality. It is therefore necessary, before going any further, to give a brief indication of the sources on which I am basing my remarks.

The Buddha, whose family name was Gotama, lived in northeastern India in the fifth century BC. His teachings are recorded in a body of scripture known to us as the Pali Canon. This, however, was written down only in the first century BC, along with a voluminous commentary. How well did three centuries of oral tradition preserve the Buddha's message? Thomas Trautmann has shown that the stories about the Buddha's family, including that of his own marriage, which appear in the commentaries must have been composed in southern India or in Sri Lanka, because the marriages are between cross-cousins, which is normal in the south but prohibited in northern India. There are other reasons too for thinking that the tradition of interpreting the canonical texts is less than perfect and that the commentators were unclear about the context in which the Buddha spoke. Yet the commentarial interpretations have been accepted by all later Buddhists, even though they are not sacred, and to a large extent by modern scholars too. We have barely begun to work, for instance, on the extent to which the Buddha's statements were intended to be taken literally (as the tradition almost always assumes) and what part was played in them by metaphor, irony and humour. This statement of the problem assumes that the oldest texts do reflect what the Buddha said; all I shall say about that here is that it is a defensible assumption and in any case a necessary one if we are to ascribe any views to the Buddha at all.

There is no *a priori* reason why I should define Buddhism as being what the Buddha meant, or even as what Buddhists have taken him to mean. But I have to define my subject matter. In the pages that follow, I am going to deal with the Buddha's own teaching as I understand it and as it was understood in the first few centuries, approximately before the turn of the Christian era. This means that I am going to leave out of account developments peculiar to the Mahāyāna. Those who regard the Pali Canon as their paramount scripture, the Buddhists in the Theravādin tradition, today live mainly in Sri Lanka and continental Southeast Asia, and their conservative interpretation for the most part goes back to those early centuries. My own interpretation overlaps with theirs but it is not always the same.

## DID THE BUDDHA
## CLAIM TO KNOW THE FUTURE?

All Buddhist traditions agree with my initial point: that the Buddha was a pragmatist who regarded his teaching as a medicine to cure suffering, the suffering of beings – all of us – caught by our own desires and delusions in the round of rebirth and redeath. The Buddha said that just as the great ocean tastes only of salt, so do his teaching (*dharma*) and monastic rules (*vinaya*) taste only of liberation. When a monk asked him some large metaphysical questions, including whether the world is finite or infinite, he refused to answer, comparing those who ask such questions to a man shot by an arrow who refuses to be treated till he knows the name of the archer. He is also supposed to have said that the things he had not taught outnumbered those he had as the leaves of a forest outnumber those one can hold in one hand.

His followers naturally came to regard the Buddha as all-wise and omniscient, and such sayings as the last one lent themselves to attributing to the Buddha complete knowledge of the future as well as of the past. For example, when many centuries later monks came to chronicle the history of Buddhism in Sri Lanka, in what is generally regarded as the first historiography in the subcontinent, they began with the Buddha on his deathbed predicting the nodal points of the tradition which finally carried Buddhism to Sri Lanka. In the oldest version he only passes those events in review in his mind. In the next and better-known chronicle, the *Mahāvaṃsa*, he actually talks to the king of the gods and predicts that his teaching will be established in Laṅkā. The brahminical chronicles, the *Purāṇas*, generally express their accounts of the past as ancient predictions, and it seems clear that Buddhist chroniclers assumed that all their history had been foreseen by the Buddha, even though they do not represent him as having gone into detail.

Already in the Canon itself there is a story of the Buddha's making a prediction about the future of his teaching – of Buddhism, we might say – and a story that is by no means implausible. The Buddha founded an institution, the monastic order, with a dual purpose: to be a community of full-time dedicated religious seeking their own salvation, and to propagate his teaching. Initially the order, the Sangha, was entirely male, but after a while the Buddha yielded to the pleas of his stepmother and founded an order for women too. The account of this event says that he did so with reluctance, and predicted that whereas otherwise his teaching would have lasted for 1,000 years, now it would last for only 500. He took it for granted that the Sangha was the vehicle necessary to preserve his teaching, and he evidently feared that the admission of women would weaken an institution founded on celibacy; he refers to this danger in another text. The tradition, however, took his words literally – or rather, as literally as was practicable. By the time of Buddhaghosa, the greatest commentator in the Theravādin tradition, the Buddha had already been dead for more than five centuries but the order of both monks and nuns was still preserving Buddhism, so the figure 500 could not stand. Buddhaghosa reinterpreted the canonical passage to make 500 mean 5,000; and ever since his day Theravāda Buddhists have believed that Buddhism will disappear from the face of the earth after 5,000 years. In my book *Precept and Practice* I have described the stages in which that disappearance is to take place. Suffice it to say that by their chronology that catastrophe is due in AD 4456. The half-way mark, as 2,500 years, was reached in 1956, and was an important date in the Theravāda Buddhist world. The date was not noted elsewhere, because other Buddhist traditions have both different dates for the Buddha and different predictive expectations of the duration of his teaching.

## CYCLICAL TIME

Though the final disappearance of Buddhism on this earth may have its spectacular concomitants – some believe that the Buddha's relics, now scattered in countless stupas, will reunite to create his body before it finally disappears – it quite lacks the finality, and hence the grandeur, of the Last Judgement. It is a whimper, not a bang. This has less to do with pessimism than with the fact that Buddhism has a cyclical view of time. Thus the 'final disappearance of Buddhism' is only its disappearance this time round. In explaining this I shall stick to our world; in some Buddhist traditions there are infinite world systems, each with its own Buddhas and Buddhisms, but that does not really affect the character of the system.

The cyclical view of time is nothing mysterious. Day and night have an obvious cyclicity, and so does the year, and several civilisations have inferred that there must also be a time cycle of a greater order of magnitude. This view is common to Buddhism, to the other heterodox religion which grew up alongside it in ancient India, namely Jainism, and to Brahminism; and we know too little of developments in the pre-literate period to say whether the pattern of thinking about cosmic time which they shared owed most to any one of these traditions or to some other tradition now lost. The Buddha does seem to have assumed a cyclicity to time, though the few statements in the Canon are not very clear and perhaps unsystematised; however, by the period of the written texts, a century or two before the turn of the Christian era, a fairly tidy system was in place and the future, in broad outline, seemed predictable – indeed, predicted. It may be significant that the very word generally used for 'future' in cosmological contexts, *anāgata*, literally means 'not (yet) come'.

This developed cosmology presents a relatively stable and predictable world, a closed (albeit potentially infinite) future. It was constructed for the most part by generations of the Buddha's early followers, and its focus of interest was not so much the physical universe, the receptacle world, as Buddhas, Buddhism and the moral condition of living beings. In the last part of this chapter I shall turn to the topic which the Buddha himself regarded as crucial, the future of the living individual, and shall there draw a starkly contrasting picture, of free wills determining open futures.

## TIME IN MYTH V. TIME IN EXPERIENCE

The Buddha explicitly denied that he was omniscient. Despite this canonical denial, his followers claimed literal omniscience for him: omniscience in the sense that he could know anything, past, present or future, by turning his mind to it. However, there has never been any suggestion of the Buddha's omnipotence. Thus it follows that the Buddha's predictions might in principle be falsified by beings exercising free will. Though logically possible, this eventuality has never been envisaged. Free will and cosmological prediction do not clash in the texts – or apparently in Buddhist minds – because they simply do not meet; they belong to different spheres of discourse.

This dual structure, like so much else in early Buddhism, can best be explained by reference to the brahminical culture which constituted its ideological background, to the speculations about Vedic sacrifice. On the one hand, this speculation embodied cosmology with its mythic time scheme; on the other, it discussed what the sacrificer was achieving by his sacrifice. Here (writes Steven Collins) 'came the idea that it was only by incessant attention to the correct maintenance of the cosmic cycle by sacrificial action that a man could produce and order a sequence of time in which to live. For Brahmanical thinking, time and continuity were not simply and deterministically given to man; rather, they are the result of a constant effort at prolongation, a constant pushing forward of life supported by the magical power of sacrifice.' Cosmic and personal time were fused in the brahmanical theory of sacrifice by the mystical identification of the sacrificer with Prajāpati, the creator god who at the same time embodied the universe. The Buddha, however, denied the validity of sacrifice and argued against (even ridiculed) the identification of the individual with the universe (microcosm with macrocosm); thus he left nothing to hold together the two concepts of time. Though Buddhists reconceptualised the spatial organisation of the universe to make it homologous with their scheme of spiritual progress, no such link was forged between cosmic time and time as we can experience it: the two topics are henceforward unconnected.

## BUDDHAS AND BODHISATTAS

To broach the topic of Buddhist cosmological prediction, I must first explain what it is that constitutes a Buddha. The word *buddha* means 'enlightened', and strictly speaking every being (normally a human being) who understands the truth which the Buddha realised and preached, the Dhamma, can also be called *buddha*. That truth is eternal, just as two times two is eternally four; but not everyone is aware of it – in fact most people most of the time are not. If we today hear and understand the Buddha's message, and realise its truth so completely that we attain enlightenment and hence freedom from the round of rebirth, we can be described as *buddha*, but *buddha* as a disciple or 'hearer', *sāvaka*: the Dhamma has been preached for us to hear. (As mentioned above, this state of affairs will last for nearly another 2,500 years – a very short period by the time-scale of Indian cosmology.)

The person whom we call a Buddha (in Pali *sammā Sambuddha*) is different in that he realises the Dhamma for himself when it is not to be heard of in this world. Not only does he realise the Dhamma unaided, he also preaches and so makes it available to others, thus re-establishing Buddhism.

I must also explain the Pali term *bodhisatta* (Sanskrit *bodhisattva*). A *bodhisatta* is a being who has taken the vow to become a Buddha. Obviously, if there are only a few such Buddhas there cannot be more than a few *bodhisattas*. Mahāyāna Buddhism is *defined* by the fact that its adherents all aspire to full Buddhahood – which is conceivable only in an infinite universe containing infinitely many Buddhas. Originally, however, the term applied retrospectively: someone who in fact became a Buddha – Gotama – had long ago vowed to do so and spent many lives accumulating the necessary qualities. Before his enlightenment Gotama was a future Buddha, a *bodhisatta*, not only from birth until that moment but for many, many previous lives. And if the attainment of Buddhahood was the result of a conscious aspiration, it should be possible to trace the moment at which that aspiration was first made. True, the impulse to seek enlightenment came at a specific moment in Gotama's early life; but we shall see that this came to be regarded as a resumption of an aspiration (like the renewal of a vow), an aspiration first made in the remote past.

There is no information in the Pali Canon about any *bodhisatta* but the one who was to become Gotama Buddha; the earlier careers of other Buddhas, past or future, are just not mentioned. There is a whole book of the Pali Canon, the *Jātaka*, which tells stories of the former lives of the Bodhisatta, that is, of the future Gotama Buddha. But in what is generally regarded as the oldest kernel of the Canon, which consists of the four large collections of sermons, of the monastic rules with amplificatory comment, and of some religious poetry, there is very little information about the Bodhisatta before he became Gotama Buddha. Not only are there very few specific mentions of his former lives; there is not even anything approaching a coherent account of the thirty-five years of his final life before he attained enlightenment.

In an article published in 1980 I collected and discussed the references in the Canon to former Buddhas. They too are sparse and mostly uninformative. I suggested that Buddhism may have adopted the idea of former Buddhas in order to cope with the problem of the Buddha's originality. Unlike the brahmin teachers, he had no lineage of gurus behind him but had on the contrary made clear that he had found his teachers inadequate. This raised a problem for the authority of his claims, and I suggested that that authority was bolstered by claiming, as the Jains did and had probably done first, that the Buddha's teaching stood in the tradition of a great line of teachers stretching back to time immemorial. So far as I am aware, no one has either approved or criticised my suggestion (so lively is intellectual debate in Buddhist studies). But the facts do seem puzzling. On the one hand, I know of no passage in the Canon which clearly indicates that the Buddha thought himself unique and unprecedented, in other words that he had *no* theory of former Buddhas. On the other hand, the older parts of the Canon (as defined above) have so little to say about those former Buddhas. It looks as if the Buddha only used the idea in a light-hearted way, not meaning it to be taken literally, or else as if the idea developed among his followers, whether during his life or after his death.

The main sermon-story about former Buddhas in the Pali Canon is called the *Mahāpadāna Sutta*. It contains accounts of the early lives, from conception to founding an order of monks, of the six Buddhas before Gotama. The story is told in full of the first of these six, Vipassi Buddha, and we are then told that the stories of the next five Buddhas (those between Vipassi and Gotama) are the same except for a few inessential details – figures and proper names – which are duly recorded. The story from conception to enlightenment centres on the famous seeing of the four signs: the old man, the sick man, the corpse and the calm renunciate. This is the episode which leads each *bodhisatta* in turn in his final life to renounce the luxuries of court life for a spiritual quest. For the past two thousand years or so this whole story has been known as the story of Prince Siddhattha, who became Gotama Buddha; but in the Canon the story is nowhere told of Siddhattha (Sanskrit Siddhārtha) – even that personal name is post-canonical. Nevertheless, the point of the *Mahāpadāna Sutta* is that the life and career of every Buddha follow the same pattern – from which we can infer that the story told of Vipassi was already held to be the story of the early life of Gotama Buddha himself.

The story of Vipassi (and of the next five Buddhas) does not contain every episode of the later classic biography. But it does contain a prediction. After the baby's birth, which is attended with miraculous auspicious portents, the king summons brahmin soothsayers, and they predict that the baby will grow up to be either a world emperor or a Buddha. It is to make sure that he does not renounce the world but becomes an emperor that his father the king shields him from the sight of human suffering, so that his encounters with ageing, sickness and death finally come as such a shock to him that they turn the course of his life. The later version of the biography, as told about Siddhattha, has one brahmin, the most perceptive, append to the prediction by the rest that the baby will definitely become a Buddha – which rather spoils the point of the story. That later version also contains a parallel episode in which a wise seer called Asita (or Kāḷa

*Figure 1* Every Buddha's life has the same pattern. Here the next Buddha, Metteyya, is a baby and a sage predicts his Buddhahood, as Asita did to Gotama's father. Wall painting, Poddalgoḍa Vihāra, Kandy District, Sri Lanka, *c.* 1950

Devala) comes to inspect the baby and first laughs, then cries; he explains that he laughs with delight because this baby will become a Buddha, but cries because by that time he will have died and so not be able to hear him preach. Asita has been compared to Simeon in the gospel of St Luke. His story occurs in the Pali Canon in what appears to be an old poem, and has then later been fitted into the biography. These stories about the Bodhisatta's infancy are of predictions; but the predictions of Buddhahood are made by holy men who have the power to see the future, but are not strictly speaking Buddhists themselves.

154

## MORAL DECLINE – AND RESURGENCE

Predictions of Buddhahood become completely integrated into Buddhism only at a later stage in the development of the doctrine of former Buddhas. Before coming to that, however, we must consider the one text in the Pali Canon which really deals with the future, and indeed is cast in an almost apocalyptic mould. Its title is the *Cakkavatti-sīhanāda Sutta*, 'The Lion Roar of the World Emperor'. A scholar may have doubts about its authenticity as a sermon of the Buddha's, at least in its present form. The Buddha's sermons are generally presented in the Canon as preached in response to a question or in some other appropriate context, whereas this text has a beginning and an ending in which the Buddha is talking to monks about something totally different: the frame simply does not fit the picture, and seems to have been taken from elsewhere. However, the Buddhist tradition accepts the text as authentic, so such scholarly scruples as to just when it originated are of secondary importance.

The myth begins with an emperor called Dalhanemi, who rules the world righteously without the need for force. A wheel-like comet hangs at a point in the sky. When the comet falls lower, he concludes that his life is drawing to a close; he hands over his kingdom to the eldest of his thousand sons and becomes a renunciate. But a week later the comet disappears. The new emperor consults his father, who explains to him that the comet cannot be passed on as an inheritance but is earned by proper rule: he must keep law and order, give wealth to the poor, and regularly take advice from wise brahmins and renunciates. The son follows his father's advice, with the result that his armies progress in every direction unimpeded; local kings voluntarily submit to him and he instructs them in personal morality, using the usual Buddhist formulae. As a result he in turn reigns happily for thousands of years, till the declination of the comet shows him that it is time for him to retire and hand over to his son. But in the third generation things go wrong. The new king keeps law and order but fails to give wealth to the poor. So theft first appears. Apprehended, the thief pleads necessity, so the king gives him money. Naturally, this encourages more stealing, and at the third case the king decides to change tactics and has the thief executed. But this leads from bad to worse. Those arrested deny the charges, and so lying begins. The execution provokes counter-violence as the population takes up arms, and violent crime proliferates.

The decline in morality leads to a decline in both lifespan (*āyus*) and good looks. Life expectancy was eighty thousand years, but in the next generation it is down to forty thousand, and it goes on halving. (All Indians believed in a natural lifespan with which one is born; one lives out that time if no disaster intervenes.) Slander then appears, as one man accuses another of theft. As good looks diminish, inequality in that respect leads to adultery – not, as one might expect, because the handsome men exploit their comparative advantage but because the uglier ones are jealous of the handsome ones and revenge themselves in that way. The lifespan constantly diminishes, more and more vices appear, and the text keeps hammering home the message that all this downward spiral was started by failure to give wealth to the poor. Mankind declines to its present condition, in which the lifespan is a hundred years.

At this point the text switches into the future. A time will come when the maximum lifespan is ten years and puberty arrives at five. At that time all the nice tastes, like butter, honey and salt, will disappear. All good deeds, and the very concept of good, will be lost. People with no shame or respect will then be praised and honoured as their opposites are now. Men will not even respect a woman on the grounds that she is mother or aunt or teacher's wife, but will behave like animals. Everyone will hate and attack everyone else, even their own parents, children and siblings. Then finally the nadir arrives. For a week everyone will perceive everyone else as a wild animal; they will all be armed and go around killing each other. But some of them will run away and hide in the mountains and forests, and live there for a week on the roots and fruit. At the end of the week those survivors will emerge. They will embrace each other, congregate, sing and congratulate each other on having survived. And they will reflect that it is wickedness which brought them so low as to kill their relatives. At this they will resolve to abstain from killing.

So the vicious cycle is set into reverse. Good looks and lifespan will increase as succeeding generations, perceiving that virtue is the best policy, gradually abandon all the vices. (One might call this a Buddhist version of social Darwinism.) Finally, the lifespan will climb back up to 80,000 years and the age of puberty to 500. There will be only three afflictions left: desire, hunger and old age. Oddly, to our way of thinking about ideal conditions, India will be very densely populated. Benares will be called Ketumatī and an emperor will rule there; his rule is described in the same terms as Daḷhanemi's at the beginning. At that time a Buddha called Metteyya will be born and attain Enlightenment. The narrator, the Buddha, says that Metteyya will realise the truth, preach it and gather a community of monks exactly as he has done himself. The emperor, after acts of great munificence, will join the Sangha under Metteyya and in due course attain nirvana.

## THE NEXT BUDDHA

That is all that we are told about the future Buddha Metteyya. His name means Kindness, and in Sanskrit has the forms Maitrī and Maitreya. The text about him covers less than a page and consists entirely of clichés normally applied to Gotama Buddha. Some centuries later a post-canonical Pali text about Metteyya was composed. It is called the *Anāgata-vaṃsa*, 'The Chronicle of the Future'. It merely extrapolates to the future Buddha the pattern of Buddhas' lives set down in the *Mahāpadāna Sutta*, so that the same events occur again but with a different set of proper names.

Though the *Cakkavatti-sīhanāda Sutta* does not put it in these terms, it seems to describe a full cycle, with the human lifespan starting at a maximum of 80,000 years, going down to 10, and then back up to 80,000 again. The implication is that the prediction of the future too is really an account of the past, for there is a set pattern in which only the names change.

How does prediction of the decline of Buddhism fit the prediction in the *Cakkavatti-sīhanāda Sutta*? As a historian, I would say that the two originate in different contexts and different types of text; the former, even though it has an eschatological tinge, concerns human history, while the latter is presented as a didactic myth. That, however, is not how Buddhists have seen it. They have fastened on to the different timescales in the two accounts and fitted the one into the other: the *Cakkavatti-sīhanāda Sutta* is on a grander scale than the decline of Buddhism, which 'The Chronicle of the Future' fits into the canonical story. This technique of encapsulation is applied not only to these two cases but also to the stray and apparently uncoordinated remarks in the Canon about cycles of the receptacle world: one kind of cycle is fitted within another, so that one reaches the kind of mind-boggling timescales which we also find in modern astronomy; in fact one type of Buddhist eon is call an 'uncountable' (*asaṃkheyya*).

Buddhas, however, are not regularly spaced through time. It seems arbitrary, but the *Mahāpadāna Sutta* says that Vipassi lived ninety-one eons ago (this kind of eon is a *kappa*, a smaller unit than an 'uncountable'); the next two Buddhas lived thirty-one eons ago; and the next three have been in this present eon, in which we have also recently had Gotama Buddha. Metteyya too will be in this eon. After him there will be an 'uncountable' eon empty of Buddhas.

Metteyya Buddha is colourless; but that is not to say that he is unimportant to Buddhists, and in fact a recent volume of scholarly essays, *Maitreya, the Future Buddha*, has been devoted to him. Most of the volume deals with the Far East and very little indeed with the Theravādin tradition. There are, however, depictions of Maitrī Buddha in many Buddhist temples in Sri Lanka; the commonest traditional way of showing him is to paint him seated cross-legged up in the Tusita heaven, where like every other Buddha in the series he spends his last life before being finally born on this earth. He is there in the Tusita heaven, the 'Heaven of Delight', right now, biding his time. He is depicted in divine regal splendour, for he is not yet a Buddha and so has not renounced splendour for the last time. Nowadays Maitrī is also represented by a statue, usually in standing position.

*Figure 2* Metteyya, the next Buddha, in the 'Heaven of Delight', awaiting his final rebirth on earth. Painting on ceiling, Dehipā-goḍa Vihāra, Kandy District, Sri Lanka, *c.* 1915. The painter is probably D. S. Mohandiram, Ananda Coomaraswamy's main informant for his classic *Mediaeval Sinhalese art*

Metteyya also figures in the Theravādin liturgy which regularly accompanies an act of merit, such as a religious donation: the meritorious person makes a wish to be reborn in the lifetime of Maitrī, so that by listening to his preaching he or she may attain nirvana. Some modern Buddhists disapprove of this wish; they point out that we are still living at a time when the teaching of Gotama Buddha is available, so that there is no need to wait. However, there is a

*Figure 3* Nātha, a Sinhalese god whom modern Sinhalese commonly identify with Metteyya and often confused with him iconographically. Painted statue, Sapugaskanda Vihāra, Colombo District, Sri Lanka. Late nineteenth century

Theravādin tradition that the last person to attain enlightenment did so in the second century BC (he was called Maliyadeva) and that the present dispensation is so far in decline that we can now only accumulate good karma in the hope of being reborn as humans in the time of Maitrī. Though his teaching lies far in the future, the vast void

160

which will follow him means that psychologically he represents a kind of last chance of salvation. A last chance, but also a second chance; it contrasts with the Christian teaching that our only hope of salvation lies in the present life.

## PRAYER AND ASPIRATION

From the doctrinal point of view, the wish to be reborn under Metteyya's dispensation is an aspiration, a kind of vow, which can only be fulfilled by the aspirant's own moral effort. To most Buddhists, however, it feels more like a prayer, a wish that will be granted as a reward for good behaviour (especially good behaviour towards Buddhist monks and nuns and sacralia). The same ambiguity surrounds what seems to me to be the one truly Buddhist type of prediction, the prediction which a Buddha is held to make that someone he encounters will in turn become a Buddha, in other words that he (it is always a male) is a *bodhisatta*. Here the doctrine of former Buddhas joins up with that of *bodhisattas*. The connection seems to have been made in the first centuries after the Buddha's death; it first appears in what are obviously among the last texts to have been admitted to the Pali Canon.

The doctrine is this. A virtuous man (or noble beast – it is not always a human being) encounters a living Buddha and performs some generous action towards him. On that occasion he makes an aspiration to become a Buddha himself. One might say that this aspiration is no different in kind from that of the ordinary worshipper who aspires to be reborn under Maitrī; it is merely different in degree. The Buddha at whose feet this aspiration is made is able to read thoughts and see, or deduce, the future; he predicts that the aspiration will in fact succeed. The ambiguity lies not in the doctrine but in the feel of the episode. It feels as if the aspirant is asking something of the current Buddha and the Buddha is granting it. In the early Indian Mahāyāna this overtone is articulated into a doctrine when the current Buddha is said to give a ritual empowerment (like anointing in the western tradition) called an *abhiṣeka* to the future

Buddha at his feet. This kind of initiation is certainly derived from Hinduism and is alien to the Theravādin tradition.

I mentioned at the outset that for Buddhists the world has no beginning. Only six Buddhas before Gotama are mentioned in the Canon, and the earliest of these, Vipassi, is said to have lived ninety-one eons ago; but nowhere is it said that he was the first Buddha, and once one admits a sequence of Buddhas there is no logical reason why there should have been a first. There is, nevertheless, in the Theravādin tradition a standard set of twenty-four Buddhas, with Gotama counting as the twenty-fifth. The *raison d'être* for this set is that under each of them the future Buddha made an aspiration for Buddhahood. He made the initial aspiration, that which launched him on his career as a *bodhisatta*, at the feet of Dīpaṃkara Buddha. At that time the future Gotama Buddha was a brahmin ascetic called Sumedha. While meditating one day he realised that Dīpaṃkara and

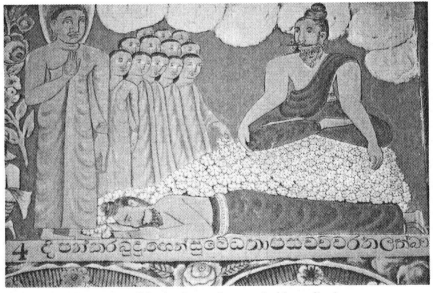

*Figure 4* Sumedha, the future Gotama, in his meditation sees Dīpaṃkara, the Buddha of his day, approaching a puddle, and prostrates himself to be walked over, aspiring thus to earn Buddhahood. Wall painting, Telambugala Vihāra, Kandy District, Sri Lanka, *c.* 1900

his monastic entourage were coming to a certain city for alms. The citizens were preparing the road. Sumedha flew over to them and volunteered to be responsible for a stretch. However, before he could fill in a big puddle, Dīpaṃkara arrived. Sumedha thereupon prostrated himself full length in the puddle so that Dīpaṃkara and his monks could walk over him without getting wet; as he did so, he thought that he did not mind if the act cost him his life, but he aspired to become a Buddha himself. Dīpaṃkara knew what he was thinking and predicted that he would become a Buddha called Gotama.

The other twenty-three Buddhas in the series are those at whose feet the future Gotama in various lives repeated that aspiration and who predicted its fulfilment. In very few cases is there any further story attached to the event. The representations of the twenty-four Buddhas in Sri Lankan temples, whether painted or sculpted, are stereotyped; the figures are differentiated only by the different forms in which the future Gotama worships at their feet. What is

*Figure 5* Part of the set of twenty-four Buddhas (numbers seven to eight). Wall painting, Uḍavela Vihāra, Kandy District, Sri Lanka, *c.* 1910

considered important is the predictions: the set of figures is indeed known in Sinhala as 'the twenty-four predictions' (*sūvisi vivaraṇa*).

Even this kind of prediction, one could argue, is a kind of spiritual prognosis. The word *vivaraṇa* does not of itself imply the future; literally it means 'revelation' — here a revelation of what is already in a sense determined. Buddhist orthodoxy would add that that determination lies in the mind of the future Buddha. This, for Buddhist doctrine, is the crux of the matter. Let me now move down from the grandiose scale of cosmology and myth to the scale on which we normally live our lives.

## SPIRITUAL PROGNOSIS
## AND THE ESCAPE FROM TIME

In the famous canonical account of the Buddha's last days, or rather last few weeks, his attendant Ānanda mentions various followers who have recently died and asks him where they have been reborn. The Buddha tells him the degree of moral progress each had made and the resultant prospects of each; he then adds that to ask such questions is fatuous, since one can judge by one's own moral condition what one's future is likely to be. What the Buddha says Ānanda could judge for himself is rather less specific than the information that Ānanda wanted about other people, and the tone of the Buddha's reply suggests that it contains a rebuke for idle curiosity. It is, however, clear in the Canon that any enlightened person can see (by clairvoyance) the rebirths of other beings, and also that such a person can judge the moral condition of another well enough to make a prognosis about how many lives it will take them to reach nirvana — a prognosis rather less precise but not so different in character from the prognosis we have seen Buddhas make for *bodhisattas*. Does this so close the future as to raise a problem for the Buddhist doctrine of free will? I think not.

Free will is the foundation of Buddhism. Good karma will bring its agent good results and bad karma bad results, and the quality of karma, though the word originally means 'act', lies purely in the intention behind it. Every sentient being has the power, and indeed the responsibility, constantly to choose between right and wrong. That is not to say that the scales are evenly balanced. Buddhism teaches what we call moral character, which is the same as moral habit, habit which may be accumulated over infinitely many lives. If one has decided many times, for instance, to kill for food, one acquires the propensity to make that decision. That may lead at death to being reborn as a tiger, a form of life in which the propensity to make that decision is extremely marked. Bad moral habits are hard to shake off, and become ever harder to abjure as one goes down the scale of life. (We can think of addiction, say to alcohol, as a model here.) By the same token, it is easier for a rich man to be honest than it is for a poor man; and the rich man was probably born into a well-to-do family, or had the chance to acquire wealth, because he was honest in former lives.

We carry the burden of our pasts, but we also create our futures. How does this fit the other Buddhist doctrine that our aim must be to attain nirvana, which is to have no future at all? How do Buddhist ethics fit into Buddhist soteriology?

To attain nirvana is to get rid of all desire and all delusion. It is to get rid of selfishness and a sense of self. The ethos and the concept are interdependent, for if one is freed from all sense of self, how could one be selfish?

I said at the outset that Buddhism is reductionist. The self is held to be no more than the label we attach to a series of physical and mental states, or better events – a series which extends over many lives. For this series, or a segment of it, a name like 'Richard Gombrich' is just a label, the self a superimposed concept, a mere construction. Long before Derrida, the Buddha was the great deconstructionist.

Thus deconstructed, we can eliminate desire, both positive and negative. The gross form of desire is the desire for externals, for money or for praise or to avoid scandal; but desire also has subtler forms. The subtlest is that which motors our continuous rebirth: the desire to continue in existence. To have this desire is to construct one's future, in the barest sense of creating time as we experience it subjectively. Thus for Buddhists it is not only the quality of our time that we are responsible for; it is our experience of time itself.

The term nirvana can be used to refer to two states, which are clearly differentiated in Buddhism but often get confused by expositors. Let me call them nirvana in life and nirvana after life. To extinguish desire and delusion is to attain nirvana in life. The body carries a certain impetus, the life-force with which one has been born and certain natural functions, such as perceptions and consciousness, which go with that; all that is part of what has been constructed by desire and ignorance in the past. The enlightened one no longer constructs, but goes on living until his or her lifespan just runs out. In other words, having attained nirvana in life one lives without desire either to have experiences or to stop them. One no longer experiences time as we unenlightened people do. I could express this by saying that Buddhism has a reductionist view of time as well as of persons: it reduces time to a relationship between acts, so that one who has lost a sense of agency has lost a sense of time.

Nirvana in life and after life is thus the same, in that it is timeless. It makes little sense to ask whether an enlightened one exists after the death of the body, because at the moment of attaining nirvana in life she ceased to construct the self, and hence its before and after. If you ask me what it feels like not to experience time, I have to give a dual answer. At one level the experience is trivially common, not merely in sleep but also when we are truly absorbed in whatever we are doing. However, such experiences are temporary: we return from them to awareness of time. We may not be able to imagine what it would be like to have lost experience of time for good, but that does not prove that it cannot happen: I cannot imagine my own death, but it will happen. What we do know is that loss of a sense of time is a private experience, and language, being social, is incapable of giving it adequate expression. Thus it is easy to understand why the experience of nirvana is mainly described in negative terms.

From what I have read about enlightened Buddhists, I would in fact doubt whether nirvana in life is really a permanent condition which lasts without interruption till nirvana after life, whether the experience of timelessness endures throughout the continued experience of social living. The more plausible claim may be that a person who has once experienced nirvana can always recreate that experience – and in particular does so at death. But this is idle speculation by someone who has never experienced it at all.

Time the scientist can measure, eternity the religious person may experience. Indian theorising on religion is wise here, for it remains aware that systematic exposition (*śāstra*), like our western theology, can never do more than complement religious experience (*yoga*). In Buddhism this distinction came to be expressed by talk of two truths, conventional and ultimate. In the Mahāyāna formulation, ultimate truth is beyond language; it does not falsify or contradict conventional truth but is on quite another plane, the plane of enlightened realisation.

How open is the future? The answer to this question is threefold. Buddhism sees the future of the receptacle world as largely determined, even predicted, but irrelevant to what should most concern us. The future of living beings is open; so much the observer can argue and anyone can experience. Both these statements about the future are, for the Mahāyāna, conventional truths. But for the enlightened, past and future are abolished, there is no time and hence no prediction. At the heart of religion lies not a theology but experience, the experience of suffering and, for some, the experience of salvation.

## FURTHER READING

Collins, Steven, *Selfless persons*, Cambridge: Cambridge University Press, 1982.

Gombrich, Richard F., *Buddhist precept and practice*, Delhi: Motilal Banarsidass, 1991 (corrected edition of *Precept and practice*, Oxford: Oxford University Press, 1971).

Gombrich, Richard F., 'The significance of former Buddhas in the Theravādin tradition', in Somaratna Balasooriya *et al.* (eds.), *Buddhist studies in honour of Walpola Rahula*, London: Gordon Fraser, 1982, 62–72.

Rhys Davids, T. W., and Rhys Davids, C. A. F. (trans.), *Dialogues of the Buddha, Part III*, London: Pali Text Society, 1921 (many times reprinted).

Saddhatissa, H. (ed. and trans.), *The birth-stories of the ten Bodhisattas and the Dasabodhisattuppattikathā*, London: Pali Text Society, 1975.

Sponberg, Alan, and Hardacre, Helen (eds.), *Maitreya, the future Buddha*, Cambridge: Cambridge University Press, 1988 (especially the articles by Jan Nattier and by Padmanabh S. Jaini).

# 8

## The last judgement

*DON CUPITT*

In the year 1960, as a newly ordained clergyman in the north of England, I attended a most unusual deathbed. The person in question was a redoubtable old matriarch who had evidently decided to die in the grand manner. As her end approached she sent out messages summoning her lifelong enemies to visit her. They were very numerous. One by one they arrived at her bedside to forgive and be forgiven. After a few days of this she was reconciled to them all, and went off to meet her maker with her conscience clear and her moral affairs in order.

It was a most splendid death, but I knew at the time that I should not see its like again. Nowadays we may indeed put our financial affairs in order and we may obtain a purely private absolution from a priest, but dying is no longer a great public performance at which we settle our moral accounts in preparation for their audit by God. At the period of which I am speaking, however, the Second World War was still not very distant. Evil had been abroad and catastrophic events had taken place. The fact that so many had died unjustly and unprepared had produced a certain revival of ancient beliefs. For example, in the 1946 film *A Matter of Life and Death*, made by Michael Powell and Emeric Pressburger, a brain-damaged pilot, hovering on the brink of death, finds himself facing a heavenly tribunal. In the 1940s, indeed, even children could still be assumed to know about the Judgement which the soul must face after death. Connoisseurs of early animated cartoons will recall *Pluto's Judgement Day* from the Walt Disney studios, in which Pluto is sentenced to hellfire by a court crowded with implacable cats, while in Hanna and Barbera's *Heavenly Puss* (1949) Tom the cat dreams that he dies but finds himself excluded from heaven until he can obtain a certificate of forgiveness from Jerry mouse. Thus, less than half a century ago, belief in the Last Judgement still had a wide popular currency. One cannot say as much today. Opinion polls report that something approaching half the population of Britain claim to believe in heaven and a smaller number, about a quarter, to believe in hell. But these beliefs no longer seem to be very vivid. If people think of hell, they think of it as a suitable destination for other people rather than for themselves. Nobody fears personal damnation any more. So far as

*Figure 1* The earliest images of the judgement of the dead come from Ancient Egypt. Here, in the presence of Horus, Anubis weighs a dead man's soul, and ibis-headed Thoth records the result

oneself is concerned, the fear of hell has declined until it is merely the fear of death. As for the Last Judgement, it has been relegated to the outermost fringes of faith. On the day I write (3 January 1991) a *Daily Telegraph* cartoon has an unshaven sandwich-board man sporting the familiar warning: 'The End is Nigh'. That is about all that we have left of the old belief. It has lost its terrors.

## A MORAL WORLD-ORDER

In the past, matters were far otherwise. Except in times of social breakdown and amongst a few philosophical sceptics, most people in most cultures have always thought that there is a rigorous moral world-order. They have thought that the world itself is so constituted quite independently of human beliefs and human social institutions, that in the long run you are going to have to pay in full for every last one of your misdeeds. The conviction that there is a moral providence has been just about universal. Dark though the future may be to us, we can therefore at least be certain of some conditional predictions of an ethical sort. If you do so-and-so, if you follow such-and-such a path of life, then you may be sure that in the long run such-and-such consequences to yourself will follow. In the long run, the moral order is infallibly upheld. God, as they used to say, is not mocked.

## A MORAL JUDGEMENT OF THE DEAD

If we now narrow the proposition down a little and ask when people began to believe that after their deaths they must face a moral tribunal, before which they must render an account of their deeds, then the answer is that this belief is first recorded in the Egyptian *Pyramid Texts* of about 2500 BC. It is subsequently found in almost every major culture-area and religious tradition, with remarkably similar imagery. The dead must cross a cosmic bridge between the worlds, their souls must be weighed in the balance, they must appear before a court presided over by a king or a god, and there is much emphasis on writing. There may be a book in which the evidence

*Figure 2* The Hereford *mappa mundi* sets the whole world under the judgement seat of Christ. To move him to pity, his mother bares her breasts

against you is already recorded; and there may be a second book in which you have been listed either for conviction or acquittal. The guilty are handed over to demons who take them down to hell for torment, while the innocent are led up a ladder or staircase to heaven.

The whole scene of the Last Judgement is so familiar, and so easily recognisable around the world, that it may represent the most universal of all religious beliefs. In particular, there is a unanimity and vividness in the way people have pictured the torments of the damned which reminds us of just how terrifying religion once was. In the Latin Christian Middle Ages the Virgin Mary was the great intercessor on behalf of guilty suffering humanity. The Hereford Cathedral *mappa mundi* puts the whole world under an image of the Last Judgement, and it sees Christ the Judge as such a fearsomely wrathful figure that his mother bares her breasts to him in the hope of awakening his compassion. The picture includes the medieval equivalent of a speech-bubble, in which she appeals to him:

> Regard, my Son, the flesh of which Thou'rt made:
> Behold the breasts on which Thou once wast laid . . .

There are some exceptions to the universality of belief in a single great Last Judgement after death. For early Yahwism in ancient Israel, for the Homeric heroes and for Confucius, religious and moral interest was concentrated upon this present world. The proud, the rebellious of heart and the transgressors would certainly be punished, but it was thought that retribution would fall upon them in this life rather than being saved up until a great Last Assize after their deaths. There is also a second group of exceptions, for there are a number of Indian and Orphic-Pythagorean traditions which believe in *karma* and rebirth. Here again you are still made to pay for your misdeeds, but you do so by carrying their consequences with you continually and from lifetime to lifetime, rather than have them reserved for a Final Court to deal with all at once.

In spite of this, even India is not without the idea of a judgement of the dead. It occurs in early Vedic times, and has tended to recur ever since. Indian cosmologies have many heavens and hells, and vivid depictions of post-mortem Judgement and hell are found in the religious art of Tibet, China, Japan and other Buddhist countries. The dead are seen as being temporarily assigned to a heaven or hell, where they enjoy or suffer the consequences of their deeds while they await their next incarnation.

In summary, we are justified in saying that for as far back as we have written and pictorial records, nearly all human beings have been convinced that there is a moral providence in the world which ensures that we will eventually get precisely what we deserve, for good or ill. In addition, belief in a final divine judgement of the dead has been very widely held. And so we arrive at our problem. The extraordinary liveliness and long prevalence of belief in divine judgement needs to be explained. What was it all about, and why has it so suddenly broken down and vanished?

*Figure 3* The Zoroastrian judgement of the dead: souls must cross the bridge between the worlds. From it the wicked fall into the flames

*Figure 4* The Muslim Judgement is rather similar to the Christian. Here, in a manuscript from Egypt or Syria, the Archangel Gabriel blows the last trump

## A DISCIPLINARY VISION OF THE WORLD

As it now appears in retrospect, the view of the world that prevailed until quite recently was highly disciplinary. It was thought that, even prior to the development of our human societies and legal systems, the world had already been designed expressly to operate as a school of moral training. This implies a thoroughly objectivist view of morality: people really thought that there is only one true and unchanging morality out there, that everything conspires to instil it, and that the Last Judgement will be a great final examination in it, which every human being must attend.

Until not long along ago the vision of the world inculcated by almost all societies was penal-moralistic, in a way that now seems strange. For we have suddenly become very different in our outlook. Amongst us morality and religion are both seen as being plural, untidy, sharply contested and everchanging. As we see it, moral values and religious beliefs are embedded in our language and culture and human institutions, and therefore cannot help but be caught up in the flux of historical change. We are not in the least surprised to find that the values and beliefs of different generations differ in exactly the measure that their manners and dress differ. Recognising all this, I am led to view my moral values and religious beliefs as being rather like my political convictions: I am reappraising them and changing them all the time in response to changing conditions. Whereas in the past people often equated rationality with adherence to timeless truth, in our historical age rationality calls for nomadism and improvisation. Although they describe themselves as 'Conservatives', Tory politicians in Britain do not see any great need to define and prove their own loyalty to an unchanging creed or timeless essence of Toryism. On the contrary, political parties nowadays are strikingly pragmatic and pride themselves on their capacity to keep on reinventing themselves in response to historical change. Similarly, in morality and religion it now appears that traditions maintain themselves only by continuously adapting themselves and reimagining themselves. Our eternal verities need constant refabricating.

You may detest all these ideas. You may remain more than half-committed to the old objectivism or realism about truths and values. But you must admit that the views I have described are very much around nowadays, and that they make the older cosmologies suddenly seem very queer indeed. What was the point of the penal and moralistic vision of the world that culminated in the Last Judgement?

## EXPLAINING THE LAST JUDGEMENT

**Law, threat and promise** We begin with this observation: in the oldest and best religious literatures available to us, statements about the future are couched in the literary forms of threat and promise.

Early human beings did not have the concept of nature as a more or less autonomous and impersonal machine that runs according to mathematical laws. Our science-based notions of prediction and forecast were therefore unknown to them. To reduce the terrors and uncertainties of the future, you had nothing to rely upon except your social institutions and the undertakings of powerful and reliable persons. Hence the enormous importance of the vocabulary of contract, vow, covenant, pledge, resolution, promise, and so forth. By rigorously binding our future behaviour towards each other, so that we know roughly what to expect of each other, we can make our life together tolerable. The future course of events becomes in some degree predictable, and it is then possible to look ahead with some confidence. The future thus begins as the product of a social contract. God is party to this contract, and he graciously extends it to the cosmic level. That is, the order and predictability of nature is the result of certain undertakings of a social kind, linguistic acts that God has performed. God's word creates the world, and he has given specific assurances that while the earth endures he will maintain the cycle of the seasons and the fertility of living things. In this way natural or physical law is seen as an extension of the civil law, and, like it, has religious significance. Cosmic and civil law-abidingness are interwoven, and both are necessary conditions of the good life. For example, a civilisation must have a calendar and a regular agricultural surplus to support the city. Its religious thought must therefore project a vision of cosmic order, and harmonise the social order with it. Ancient thought thus shows rather clearly the extent to which reality is a social, and indeed an ethical, construction; by which I mean that the extent to which we can recognise a reliable order of things in the cosmos and in society is a function of the strength of our social discipline. In the long run everything depends upon law and sanctions. Human beings were taught constancy in their social behaviour and their view of the world by long subjection to closely defined and stringently enforced legal systems, and by carefully graduated but always highly physical punishments.

These considerations, I suggest, throw some light on the idea of the

Last Judgement. They help to explain why it was a legal and not just a moral inquisition. It was a court hearing presided over by a royal Judge, and its proceedings were in the very highest degree public and conclusive. We tend nowadays to forget that the word 'apocalypse' just means open publication or disclosure. The Last Judgement is the ferreting out of every secret, and the exposure to the light of every smallest sin. It symbolises the way the penal system, rigorously enforced, makes life make sense. On the Last Day the final Truth, the final moral accounting and the final ordering of reality fully coincide with each other. The Last Judgement is the complete triumph of the principle of law and order, and therefore the ultimate rationalisation of the world. The scenes of torment in Hell reveal the price in human suffering that has to be paid – and indeed already *has* been paid – to make possible the large-scale, prosperous and relatively peaceful bureaucratic societies of today. The scenes of peace and harmony in Heaven reveal the benefits we stand to gain.

A full history of ideas about the future has not been written, and perhaps can never be written. But I am suggesting that in the early civilisations social discipline was extremely severe. The notion that crime must inevitably be followed by highly unpleasant physical punishment was branded upon people's hides, and therefore upon their emotions. This is the source of all our ideas about the necessity with which effect must follow cause. So, for most people and for thousands of years, the single most vivid thought about the future that they had was the terrifying certainty that all of their so-far-undetected crimes must one day be exposed and punished. Hence the emotional force of the idea of a Last Judgement, and hence, too, the fact that everywhere hell was more vividly imagined than Heaven.

These ideas about the terror of the future, and about the movement of time towards a stupendous consummation, raise some very interesting considerations about the relation between history and sin.

**Linear time, history and sin** The earliest pre-literate and ceremonial societies did not have our conceptions of history and linear time. Authority in such societies resided with the elders, who kept recalling the young to the standards and values of the past. Everything needed a traditional warrant to make it real and effective, and cultural identity was secured by the regular repetition of myths and rituals. These ceremonies united present time with the primal time in which all the standards had been set. By this I mean that in primal time Gods, heroes and ancestors had between them created everything, set up everything and done everything exactly as it ought to be. Our task is only to keep to the paths that they have marked out for us. The movement by which in rituals time kept on returning to its own beginning was possible because, happily, the natural movement of time was not linear but cyclical. Even today the ancient cyclical conception of time is not forgotten, because so many of our basic units of time are still linked to natural rhythms such as those of the day, the month, the annual round of the seasons, the various human and animal rhythms and life-cycles, and the revolutions of the heavenly bodies. The world had been created finished and perfect in the beginning, and the function of rituals was to keep following the natural curve of present time back into primal time. Nature is able to renew itself every year, and human society must do the same. The perennial human fear of Chronos, devouring Time, was thus assuaged as everything was cleansed and reborn by regularly returning to its own origin. In our own society we can still see all this illustrated in our rites of passage, and in our annual round of religious feasts and fasts. Perhaps the most vivid examples of the reunion of present time with standard-setting and mythical past time are found in the Jewish Passover, the Christian Eucharist, and on Christmas Eve.

In such a context, then, religion seeks to overcome history by annually circling profane time back into its mythic, sacred beginning. Do we remember that the word for a year, *annus*, means a ring or circle? A thing or a deed is real and sanctioned only in so far as it is sanctified by participation in its sacred origin. That is, reality is acquired solely by repetition or participation. A person or thing becomes real only in so far as it ceases to be merely itself and becomes instead the universal; and action becomes effective in so far as we cease trying to act of ourselves, and are content instead to imitate and repeat the gestures of another. Mythical and ritual thinking strive always to eliminate that which is merely profane and fleeting – the one-off, the contingent, the merely individual and particular. The holy is always enduring and general. Everything needs to be taken up into sacred emblematic patterns that endlessly reiterate themselves.

It follows that for archaic thought, history is sin. Virtue and holiness are equated with cyclical time and the due repetition of hallowed archetypal forms. Unprecedented action is unauthorised, unhallowed and therefore sinful action. When human beings try to break out of the ritual universe and start to do their own thing in their own strength, then they fall, both into sin and into history. By their action they create non-repeating linear time, and so move steadily away from life's old origin and centre.

The historical human being seeks self-realisation as a unique individual subject through action in linear time. We become real self-made individuals, each with a unique personal history. But like Hans Christian Andersen's Little Mermaid, we pay a heavy dual price: we live conscious of being exiled from the archaic paradise of cyclical time and the endless rejuvenation of all things, and, having chosen irreversible linear time, we cannot help knowing that we are mortal sinners, heading always and only in a straight line towards suffering and death.

The historical human being's awareness of alienation, sin and death gives rise to a new kind of religion, the religion of redemption at the end of time. In this new context the symbol of the Last Judgement does two jobs. It symbolises to us our sinful mortal condition and the condemnation it merits; and it warns us to begin seeking redemption. Furthermore, it gives historical time a telos or goal, reassuring us that we are not simply rushing away from Eden into the outer darkness. Even historical time thus in a certain sense comes full circle and returns us at last to confront our own origin. The old annual rejuvenation of the world is replaced by a new hope, the hope of a once-and-for-all resurrection of the dead.

**The deferral of the end** This brings me to a third and last strand in the idea of the Last Judgement. Especially in our own western tradition, the tension of history has been maintained by the doctrine that we do not know and cannot know the precise date of the end of all things. The End is promised soon, but just how soon we are not told, and it seems to keep on being postponed. This repeated deferral maintains tension, vigilance, expectation and striving. It keeps our nerves on edge. We know we must do all that we can, while we have time. We are told to make the best use we can of our extended deadline. We are assured that history is going somewhere, that it all happens only once, and that there is a point of view from which one day it will suddenly be judged complete. It will then be abruptly wound up. Life is like a classic detective story: at its close the plot will be resolved and everything will be explained. Some of us are going to get a nasty shock. But precisely when will we find we are at the end of the book? We do not know: we are kept in suspense. There is often a scene at the end of the classic detective story where the detective arranges for all the leading characters to be assembled for the dénouement. This scene corresponds to the *Parousia*, the end of history. We have a strong interest in knowing when that scene is going to take place, for we will be involved. But we are not to know when our summons will come. Instead we have to live like the Jew in eastern Europe, who was said always to have a suitcase packed ready for flight. So we believers should all of us live as people upon whom the End may come at any moment. We cannot live as people who feel they are entitled to ten years, or thirty years, in which to get their lives into a rounded shape and ready for inspection. We may be suddenly raided and called to account at any time.

The stress of waiting for an End, the precise dating of which is entirely beyond our control but which keeps on being mysteriously deferred, has troubled the whole Western tradition. In various ways we who live in linear time have always struggled against it, but always without success. There has been a long history of prophets arising to declare that the End is now very close, and sometimes even assuring their followers that it will come on a specified day. There has also been a long history of numerological enthusiasm. Great thinkers, such as Newton, have striven precisely to calculate the date of the End. But all such prophecies and calculations are eventually falsified, and the problem of deferral returns. There have also been many, many people who have attempted by various extreme measures to force God to act, and so to hasten the End. They have tried to short-circuit history by extreme asceticism going even as far as suicide, or by extreme utopianism, as when they have set out to escape from or overthrow the existing social order and build in place of it the kingdom of God on earth. But again, all such attempts to complete history suddenly and by force themselves become in due course just part of history.

These are by no means the only devices by which people have tried to escape from the fearsome *lapse* of linear time. For country people in the Middle Ages, life was so dominated by the agricultural year and the Church's annual round of feasts as in effect to reinstate cyclical time. The monks went still further, living outside history in the antechamber of eternity. There is also a long tradition of philosophies of history, from Joachim of Fiore to Karl Marx, which postulate a great world-historical story and place the present within it. There is an element of forbidden, impossible knowledge in any such story, because it purports to let us in on the secret of exactly what story God is telling through world-history. By situating ourselves within this story we get an idea of what is going on, and even of when God may judge the moment right to bring down the curtain. We desire this bit of forbidden knowledge of what the real story is, because without it the sheer lapse of time, its constant falling away, its running out like the sand in an hour-glass, seems to portend universal annihilation. Humans hate the thought that everything is contingent and nothing is conserved for ever. The Grand Narrative (as we nowadays call it) sets out to overcome this fear by claiming that a thread of meaning runs all through time. Not everything passes away, for we can trace in history the fulfilment of a moral purpose and the realisation of values. Our own activities can be justified in so far as we are able to describe them as contributing to this larger background purposiveness that runs through all events. If so, then not everything is forgotten: on the contrary, something of what we have done, and perhaps even something of what we have made of ourselves, is remembered and conserved.

All this reminds us that the court scene at the Last Judgment was not concerned solely with the punishment of the wicked, the righting of wrongs and the balancing of moral accounts. The court was also a place of salvation. The king in giving judgement appeared like the sun at dawn, glorifying his own name by restoring the social order, just as the sun glorifies itself by its daily renewal of the cosmic order. Nowadays the acquittal of the not-guilty has a rather negative sound to it; but in early antiquity the king by vindicating the innocent caused them to shine forth like stars. In this context the Last Judgement becomes an act of final redemption, in which God vindicates himself by vindicating his righteous servants. All truth is published, the meaning of the whole cosmic story is finally made clear, and the elect gain salvation.

But this final event is delayed and delayed, and meanwhile the terrible eroding *lapse* of time continues. As succeeding generations slip away into oblivion, we seek reassurance. The philosophies of history are thus meant to encourage us and revive our fading hopes. They tell stories designed to renew our faith that world-history does have a telos. Yes, it will indeed end with a bang and not a whimper – but not just yet. And the irony of the philosophies of history is that they are caught in an unpleasant double-bind. They tell stories that encourage us to hope that world-history is getting somewhere, but only when it has actually got there can they themselves be vindicated.

To repeat and generalise the point: the story that at the end of time everything is going to make sense cannot itself fully make sense until the end of time. An apocalyptic book promising that in his own good time, and not before, God will make everything plain cannot itself jump the gun. By definition, it must wait for God's good time before it can itself become fully plain. So our anxiety and uncertainty about where it is all going and when it will end cannot be wholly allayed.

## THE LAWCOURT:
## SOCIAL ORDER, COSMIC ORDER

A number of points stand out from this analysis of the meaning of the Last Judgement. First, we can date the idea from its iconography. From the court scene, the books, the balances and the penalties it is clear that the Last Judgement is part of the ideology of the early literate civilisations. This is a world whose greatest institution is, arguably, the lawcourt. It is an arena of reason, where disputes are resolved and the social order is upheld. In the lawcourt the basic vocabulary of logic and epistemology was first developed. For example, in court people debate how correctly to characterise a particular case, and whether or not it falls under a law. And throughout our whole tradition since Aristotle we have accordingly described as 'judgement' the act of mind by which a particular is recognised as falling under a general concept. We have regarded every cognitive act and every true proposition as having this logical form. Again, it was in the lawcourt that we first developed our ideas of causality, responsibility and freedom, and of action, intention, property, person and right. In the lawcourt we worked out our ideas of what evidence is, what makes it relevant, how to evaluate it, and of how it may confirm or disprove hypotheses about what is or has been the case. The lawcourt was the nursery of reason. From it there developed the vision of a law-governed cosmos, and the use of quasi-legal methods of enquiry to discover what are the laws by which the world is ruled.

It is not surprising, then, that in ancient religious thought God was so often portrayed as permanently holding court in heaven, and as threatening or promising to hold a final and universal court at the end of the world.

## PARADOXES IN THE
## IDEA OF A MORAL PROVIDENCE

Yet notice some strange paradoxes in all this. It is said both that the world is already wholly subject to the divine will, and that, precisely because it is *not* so, we are threatened with a catastrophic divine judgement in the future. The traditional themes of Creation, Providence and Revelation portray a world already wholly subject to divine law. God's moral law is already revealed, his moral providence already operates, and already nothing can happen apart from his will. Yet it is also said that the world is in revolt against God and is becoming increasingly wicked and disorderly. On this account God's absolute sovereignty over the world will not be fully asserted until Judgement Day. So the blessed vision of a world everywhere ruled by divine law is held strangely suspended between a mythic past and a deferred future.

What, then, is its present status supposed to be? The answer seems to be that it functions as a guiding picture, an idealised memory, a promise and a threat. For consider a second paradox. Religious doctrine assures us that God's moral providence is already everywhere at work in the world to make sure that in the long run we get what we merit. Since God is by definition all-powerful, all-good and all-knowing, the penal system that he operates must already be one hundred per cent efficient. But if so, why does it need any topping up, either by the civil penal system or by a finalising Last Judgement? If God's moral providence is already fully operative, there will be nothing left over for the topping-up processes to deal with.

Our argument as a whole has shown that religion and law were in the past the chief sources of our ideas about the future. One might say that they made the future what it is today. In particular, they were concerned to stress the importance and value of cosmic and social law-abidingness. In the presentation of this theme there are always elements of threat and promise. We are warned of impending doom unless we change our ways, or we are promised harmony and well-being if we do so-and-so. We are agents and we are evaluators, so that when a hypothetical state of affairs is presented to us it must appear as either a doom to be averted or a goal to be achieved. Talk about the future is always admonitory: prophecies, predictions, scenarios and religious doctrines portray the long-term outcomes of various choices that we may make. We are pictured as standing at a crossroads: the different paths before us will take us to different destinations. The future is not already out there, laid on and fixed. It is a choice that we have to make.

## THE FUTURE IS AN ADMONITORY FICTION

So we can close with an explanation of why it is that in our tradition the Last Judgement and the End of All Things keep being postponed. It is an error, a kind of idolatry, to want your religious beliefs to be objectively true and fully actualised in present experience. Rather, they are admonitory pictures and guiding ideals. They have to stay out of reach. The postponement of the End opens up history, our human breathing-space, the period in which we sinful mortals don't have and don't know – and therefore must choose instead.

To guide our choices we paint pictures which portray the outcomes of the various options that lie before us. Such pictures, I freely acknowledge, are mere human fictions. They are by no means wholly arbitrary fictions, and it makes a great deal of difference which of them one lives by. Nevertheless, they are just fictions, and we who live after the end of the old realistic, dogmatic type of belief have to accept that. Personally, I am content with voluntary, fictional religion, linear time and an open future. Are you?

## FURTHER READING

Baudrillard, Jean, 'The year 2000 will not take place', in *FUTUR*FALL: Excursions into Post-Modernity*, Sydney: Power Institute of Fine Arts, 1986.

Brandon, S. G. F., *The judgment of the dead*, London: Weidenfeld and Nicolson, 1967.

Deleuze, Gilles, *Nietzsche and philosophy*, translated by Hugh Tomlinson, London: Athlone, 1983.

Eliade, Mircea, *The myth of the eternal return* (Bollingen series 46), London: Routledge and Kegan Paul, 1954.

# NOTES ON CONTRIBUTORS

*AVERIL CAMERON* is Professor of Late Antique and Byzantine Studies at King's College, London, and is particularly interested in cultural and intellectual history. Her most recent book is *Christianity and the rhetoric of empire* (Berkeley and Los Angeles, 1991), and she is currently writing a study of cultural change during the 'Byzantine dark age' in the seventh and eighth centuries *AD*.

*DON CUPITT*, Dean of Emmanuel College, Cambridge, is University Lecturer in the philosophy of religion. He is the author of some twenty books, including *Taking leave of God, The sea of faith, The long-legged fly* and *The time being*.

*RICHARD GOMBRICH* has been Boden Professor of Sanskrit at Oxford University since 1976. His main research interests are the history and anthropology of Buddhism. Recent books include *Theravada Buddhism: a social history* (London, 1988); *Buddhism transformed: Religious change in Sri Lanka* (Princeton, 1988, with G. Obeyesekere); and *Buddhist precept and practice* (Delhi, 1991) – a corrected edition of *Precept and practice* (Oxford, 1971).

*FRANK HAHN* is Professor of Economics in the University of Cambridge and a Fellow of Churchill College. He has honorary degrees from several other universities, is a Fellow of the British Academy, and a member of the American National Academy of Sciences. He has published numerous essays on aspects of economic theory.

*STEPHEN HAWKING* CH is Lucasian Professor of Mathematics at Cambridge, occupying the same chair that was once held by Isaac Newton. While studying for his Ph.D. at Cambridge, he was diagnosed as suffering from motor neurone disease and did not expect to live to finish his doctorate. He subsequently achieved international recognition for his work on general relativity and quantum cosmology, and public acclaim for his international bestseller *A brief history of time*. He holds twelve honorary degrees and is a Fellow of the Royal Society.

LEO HOWE, editor of the volume, lectures in social anthropology at Cambridge and is a Fellow of Darwin College. He has conducted anthropological fieldwork in Indonesia and Northern Ireland and is the author of *Being unemployed in Northern Ireland: an ethnographic study* (Cambridge, 1990).

IAN KENNEDY is Professor of Medical Law and Ethics, and Dean of the Law School, at King's College, London. He is a member of several public bodies in Britain, including the General Medical Council and the Medicines Commission. His most recent books are *Medical law: test and materials*, and *Treat me right* (Oxford, 1991).

SIMON SCHAFFER lectures in history and philosophy of science in the University of Cambridge and is a Fellow of Darwin College. He has recently edited collections of essays on the history of experimental science and on the life and work of Robert Hooke and of William Whewell.

IAN STEWART is Professor of Mathematics at the University of Warwick, where he is director of its interdisciplinary mathematical research programme. He is the author of *The problems of mathematics, Does God play dice?*, and *Game, set and math*. He writes the mathematical recreations column of *Scientific American* and appears regularly in *New Scientist*. His research area is non-linear dynamics and its applications, with an emphasis on the effects of symmetry. He is currently working on symmetric chaos and on the patterns observed in the gaits of animals and insects.

## *ACKNOWLEDGEMENTS*

### CHAPTER 1

All illustrations after originals by Stephen W. Hawking.

### CHAPTER 2

*Figure 1* NASA; *Figures 2, 3, 4, 5* Ian Stewart; *Figure 6* Harry Swinney; *Figure 7* Tom Mullin; *Figure 8* Ian Stewart; *Figure 9* Michael J. Field and Martin Golubitsky.

### CHAPTER 3

*Figure 1* French print, 1857; *Figure 2* Lithograph by Honoré Daumier, 1853; *Figure 3* Woodcut by Jiri Daschitzsky; *Figure 4* German print, late seventeenth century; *Figure 5* Bodleian Library, Oxford; *Figure 6* from Buffon, *Histoire naturelle*, Paris, 1947, p. 127; *Figure 7* from Grandville, *Un autre monde*, Paris, 1844, Chapter 15.

### CHAPTER 4

*Figures 1, 2* after originals by Frank Hahn.

### CHAPTER 6

*Figure 1* Ms Laur. Plut. IX.28, fol. 95v; *Figure 2* Paris, Musée du Louvre, Département des antiquités grecques et romaines, no. Bj 2180 (Ma'aret en-Noman treasure); *Figure 3* Rome, Biblioteca Apostolica Vaticana, Vat. gr. 699, fol. 89r; *Figure 4* Oxford, Bodleian Library, Ms Auct. F.5.4, fol. 1v; *Figure 5* London, British Library, Royal 19.A.IV. Fol. 13v.

### CHAPTER 7

All photographs by Richard Gombrich.

### CHAPTER 8

Photographs courtesy Don Cupitt, except *Figure 2* which is reproduced by kind permission of the Dean and Chapter of Hereford Cathedral.

# INDEX

Printed in the United States
25688LVS00001B/377-380